THE ENGINEERING YEARS
1963 - 1982

THE SEARCH FOR IDENTITY SERIES
– BOOK THREE

I0500214

Harry MASON

[Type text]

[Type text]

Version 1.0 – June 2017

Published by HARRY MASON at CreateSpace

ISBN: **978-1547289189**

Copyright © 2017 by HARRY MASON

This book is a memoir. Names, characters, places and incidents are products of the author's memory and/or by web investigation, any errors of fact therefore are my fault. Some names and/or incidents may have been changed to save embarrassment.

Dedicated to our lovely granddaughters Lilly Rose and Ruby Isabella for whom this book was written. With a special thanks to Carol who lived with me through this journey and always gave me more support and love than I at times deserved.

[Type text]

Table of Contents

Chapter One -AHQ Melbourne Jan 1972- Mar 1974

On completion of the long period of training the system had invested in it was time to begin to repay the debt. My posting was to Army Headquarters Melbourne in Albert Park Barracks, to the Directorate of Electrical and Mechanical Engineering (DEME) as the Staff Officer grade three in charge of the maintenance of all electronic surveying equipment. That was shortened to SO3 Survey, the grade three means it is a Captain's position, and I was promoted temporarily.

Initially the position was filled by a National Service officer and he had done some of the lead up work and I picked it up from him. I and the other SO3s reported to a Major, the SO2 Electrical. His name was Eric Lauritzen and he came up from a Craftsman, Radar to his present position and was a gentleman. He taught me well and gave me one of the most significant jobs I ever got, but more of that later.

We had our removal just after Christmas and as there was no MQ available we moved into a rental that just happened to be opposite Carol's parents place in Baileyana Street. It was a fortunate situation brought on by the house owners moving to Gove in the NT for work. It worked for them on another level as well as we looked after their two Siamese cats after they left. The trip to Albert Park in Melbourne from Frankston is over 40 kilometres and took about an hour and 5 minutes if we caught the traffic, but the convenience of having Carol's parents close at hand became obvious shortly.

The Royal Australian Survey Corps are responsible for the surveying; drafting and reproduction of all maps used by the Australian army and currently had operations ongoing in

Indonesia and on mainland Australia near the Gulf of Carpentaria.

The School of Survey was at Bonegilla on the shores of the Hume Weir where a few years earlier I had completed the diving course, and the Survey Regiment, its main unit and source of the production part of their work was in Bendigo in Victoria.

They had just taken on board a new surveying system called Aerodist which was 'survey-speak' for an airborne distance measuring system.

In 1963-64 the original Aerodist was a development from the Tellurometer system used forground measurement of geodetic distances. It was designed for simultaneous measurement, during movement, from an airborne station to two or more stationary ground stations. The major differences from the ground based Tellurometer equipment occurred at the airborne station where the directional parabolic reflector of the antenna system had been replaced by a small flat plate reflector which could be rotated relative to a fixed housing and provision had been made for continuous simultaneous recording of distance as a function of time by the use of a 3-channel chart recorder.

The Army's 1972 version added a PDP 8E computer that was intended to perform the calculations and provide for real time processing of data. The PDP 8E was a 32 bit 4.8 k ferrite cored memory computer expanded to 9.6 k for our use.

Basically a precision altimeter and a number of Tellurometers inputting direct distance measuring information into a computer, with a backup tape recorder, all mounted in a commercially contracted King Air aircraft. The idea was that the aircraft would fly along a bearing between two land based Tellurometers up to 70km either side of the bearing being flown; the Tellurometer emitted a super high frequency wave; the remote stations reradiated the incoming wave in a similar but modulated wave and the resulting phase shift was a measure of the distance travelled.

The varying distance recorded, combined with the change of elevation, would give a profile of the terrain below.

Seeing this was to be my baby it was important for me to receive a Month's training on it in North Sydney. It started with training on one of the more important elements, the DEC PDP8e computer for a fortnight from 24 January to 11 February 1972.

This was followed immediately by a fortnight from 12 February to 25 February 1972 at the prime contractor Digital Applications, learning how it was all to come together and how we, Raeme, were expected to maintain and repair when necessary.

During this series of courses I was accompanied by WO1 Artificer, Radar, John Strautins, and we were to be the support team for the Aerodist system whilst it was in service. For the next month or so John and I worked closely with the Survey regiment, where Peter Eddy my back fence neighbour from North Beach was the Operations Officer. We became de facto RA Svy Corps members and attended their operational planning conferences held in April each year for the next couple of years.

****

In the bowels of the beast that monitors and controls the movement of officers and qualification for promotion it had come to the attention of someone that I had not completed the course known as the Tactics 1 (TAC 1)for Captain which is required to be done prior to promotion to the substantive rank of Captain. Because of my extended period of LTS I had been overlooked and as I was due for promotion on time in June 1972 something had to be done quickly. What was done was they placed me on the next TAC 1 course that began on Monday 28 February 1972.

This meant that I was home for the night of 26 February only, before flying to the Jungle Training Centre, Canungra, for a three week course, from 28 February to 17 March 1972.

I am sure that all bar me on the course had served in SVN, and had years of unit service. I was a real novice. The first day they gave us an introductory test to assess our prior knowledge. It may as well have been written in Swahili for they were asking questions about rates of fire and effective and maximum ranges of weapons and weapon systems I had not even heard of in some cases.

For example when asked about the RPG 2s and RPG7s. I thought it stood for rounds per gun not rocket propelled grenade. These were the weapons of Russian design used by the Viet Cong in SVN.

During the first week we went into the hinterland jungle and did what are called Tactical Exercises Without Troops (TEWTs). For me it was all a learning experience and I learnt to determine the enemy, his location, likely weapons and intentions and develop a plan to counter him and defeat him. We looked at the Attack and Defence phases of war, in both a conventional war setting, e.g. WW2 or Korea, and counter revolutionary warfare as in SVN. The task to be completed was in two parts, the assessment of the situation and your plan to beat the enemy, and how good your orders were to the troops having to do the task. The assessment of the situation is called making an appreciation and you were marked on your appreciation just as much as your plan and how you gave your orders in the orders group (O Gp.) that followed. We did it in syndicates of four or five and I had been adopted by some of the more experienced guys and we made a close knit syndicate.

It was no surprise that I was told I had to stay in camp on the weekend, while the others enjoyed the Gold Coast, and learn all this basic stuff that had been left off my Associateship course. It was rote learning and I devoted all Saturday, day and evening, to studying it.

A couple of Arms guys, one Infantry and the other Artillery, from our syndicate came back early on the Sunday and quizzed me on the study I had done and by the end of the day I was pretty confident that the next day's test would be a significant improvement. I was able to recall all information easily and was then back on the road to a pass. Week two was similar to week one but we ventured out for a night navigation exercise on one of the darker nights, I thought so anyway, and I learnt to respect if not enjoy the clutches of foliage with names like wait-a-while with its hook shaped barbs and lantana which was thick, prickly but even worse caused a rash to develop if you got it on your exposed skin. Wandering along following my compass and trying to pierce the gloom I found myself about a metre higher than the guy alongside me.

I had walked up on to the trunk of a fallen tree and was steadily ascending into the gloom. Heaven only knows where I was going or perhaps heaven was where I was going. Depends how long the tree was, I suppose.

Now back in the good books I was allowed to go on leave to Surfers' Paradise. We stayed in a motel and had a few beers by a pool and enjoyed a Hawaiian luau. The next day we decided to go for a walk along the beach and set off from Surfers' heading south. We walked to Broadbeach, about 6 km and were all dry and sunburnt by the time we got there so had a drink and walked back in time for the ride back to JTC.

Next week was our last week and we were taken to O'Reilly's Guesthouse in the hinterland, where we lived in the tourist chalets.

Bernard O'Reilly was still alive then but an old man and it was his daughter who gave us a slide show presentation of his efforts in the 30's when he searched for and found the survivors of the Stinson air plane crash which occurred on 19 February 1937.

The Stinson Model A airliner disappeared during a flight from Brisbane to Sydney, carrying five passengers and two pilots. Both pilots and two passengers were killed in the crash. One of the surviving passengers died while attempting to bring help to the other survivors. The wreckage was found by Bernard O'Reilly who lived approximately 8 kilometres from the site and went looking for the aircraft believing it had failed to cross the border. The aircraft had crashed in the McPherson Range on the border between Queensland and New South Wales the terrain is rough and very difficult but with his horse and knowledge of the bush he succeeded when everyone else had given up.

Our task during the week was to gain experience in navigating through the rougher hilly terrain and we had a withdrawal TEWT to complete, covering the extraction of troops from a position near Python Rock back to O'Reilly's Guest house.

****

On Friday 17 March 1972 I arrived back at our rental in Baileyana Street tired with all the courses I had had but now up to date with my peers and ready for substantive promotion to Captain on 30 June 1972, and, more

11

importantly equipped to do my job. It had been over two months except for one night that we had been separated. The significance of this will become obvious later.

Because Geoff Hawker only lived down the road and also worked at DEME we shared the driving leaving very early to miss the traffic and trying to get away early in the afternoon to miss it coming back. We were doing flex time before it became fashionable.

The Fitzroy VFL Club ground was just across the road and I heard that we could use their gym when they were not using it, i.e. early in the morning.

I went over to check it out and found that now ex Wo1 Ray Keane was the head trainer. He was only too pleased to get his hands on one of his 'super-fits' as he referred to us, and used us as inspiration to the others doing the early morning classes. Some were footballers needing more and harder work, and others who varied from simply keen-to-be-fit people up to a few recovering from heart problem and using a prescribed fitness regime to get back to health.

He had a fatty's group (footballers not fit enough) that had to report very early and be taken for a 'brisk run' around the track surrounding Albert Park Lake, before being punished in the gym. Still they were earning money to play VFL.

Occasionally I would arrive early enough to be included in this elite group whether I liked it or not, we (Ray and I) never ever got over the OCS thing where he was boss and I did what I was told. It was pointless me telling him I was not on the Fitzroy payroll.

Chapter Two – Survey in Far NQ

They, RA Svy had two operations in the works for the dry season, one in New Guinea and another about to start on the Gulf of Carpentaria operating out of the airfield at Normanton using the Aerodist Equipment. John and I were told to join them in Normanton after they had been on the ground for a month or so as things were not going according to Hoyle. We flew to Cairns and overnighted at a boarding house on the Esplanade there before boarding a Bush Pilots' aircraft to take us to Normanton, where we became part of the survey team.

I was to get used to the idea of an 11x11 tent with a stretcher and a wooden box as a tables bedside, over the next year and a bit. Normanton in the dry is a dusty but pleasing enough little town with three hotels, a Burns Philip store, a post office, a school and a police station, plus the odd house. There is an aboriginal reserve (that was the correct terminology in 1972) nearby but whites were not allowed access. I remember walking into town on QLD Election voting day 27 May1972 and it was 0900hrs and I took a photo of a completely empty street scene looking from the Centre Hotel towards the Burns Philip building. There was not a soul in sight. On the way back I had to pass the Normanton hotel and looked in to check it out. it was a wooden bar top and could have been from a western Movie set except that there was a baby crocodile sitting in pride of place on the left hand edge of the bar. Of course I was fascinated and asked if I could pat it. "Sure" was the answer, "no problem." There was no problem for them or the croc but as for me the bloody thing tried to take my finger off. I was lucky and it missed.

We had set up the Aerodist and checked all the equipment and it appeared to be working on the ground giving good response to two dispersed ground stations. However when we tried to do it whilst airborne we had poor results.

The computer was not processing the inputted data as expected and it was necessary for the surveyors to break down the recorded data in the comfort of an air-conditioned cabin set up on the side of the airfield. Another cabin was used by our technicians as a workshop. After a frustrating afternoon of trying and failing to identify and fix the problem we gave up and went into town to check out the night life. This time we went to the Centre Hotel, because we had been told they had the only air conditioned bar in town, it was called the Private Lounge and was a 'whites only' area. I was about to have my naïve little city eyes opened to the realities of life in the 'Deep North'.

The private lounge was not a large room and one of the first men we met there was the local ABC rounds man who just happened to have been our landlord in Cairns. He owned the Cairns Guest House.

We went exploring and checked the hotel out and saw there was a front bar that was for the Aboriginal clientele. I believe they had a tab system where they handed in their government cheques and drew down on it for drinks and cigarettes.

No idea if it was fair or not but it worked. One little sidebar, pun intended, was the way brawls in the bar were handled. If there were two or more having a difference of opinion, *inside* the hotel, the barman would grab what looked like a truncheon and jump the bar and with a few sharp blows resolve the dispute and kick all involved out. "No more drinks tonight," he would say and that was a greater punishment than the blows sustained. To add pathos to this drama, the barman would see how many beers were knocked over when he had to jump the bar and refill them at the expense of the miscreants, rough justice indeed, but everyone seemed to accept this as the Norm in Normanton

Out the back of the hotel there were some caravan sites and a few Crocodiles in cages and wedge tailed eagles in Aviaries. I got talking to the chef who was taking a break and asked him about life in Normanton.

He waxed lyrical about the fishing down at Karumba the fishing port on the gulf and explained that trucks would drive through from Townsville load up with the frozen prawns and barramundi and drive back to Townsville so the fish could be flown on to the markets down south. I had never had Barramundi so asked him what they were like and if I could have a meal of it one night. He told me he would select a good one and if I came the next evening he would cook it especially for me. It remains the best barramundi I have ever eaten.

Our camp was supplied under contract by the local shopkeepers and the management of this aspect of our life was in the hands of a very young National Service 2Lt. He always seemed to be on top of things and we had an all ranks bar and mess on the airfield so there was no need to go into town if you did not want to. Suddenly he seemed to excel himself and we were eating prawns three times a day and as bar snacks. When asked he admitted he had lucked on to a truckie in town who had parked his truck near the pub, met a young half caste girl, and she was living with him in the truck. To pay for this extravagant lifestyle he was selling his cargo off at good prices. I have no idea if he thought he could get away with it or not but after an incident in the Private lounge the law moved him on. One night full of booze he tried to bring his girlfriend in for a drink in the cool. She was terrified of what could happen and in a short time they were out drinking in the front bar. In the morning he was run out of town and she was back on the reserve.

After another fruitless week with the Aerodist we borrowed a Landrover and drove down the sandy rough track to Karumba. There was a lodge there that catered for tourists and a good number of prawn trawlers and fishing boats. We strolled along the pristine beach and watched a couple of young kids casting a net to catch bait fish that they stored in a dinghy half full of water and sold to the tourists as bait. I enjoyed a quick swim and ignored the signs we saw later referring to Crocodiles.

Back at camp we only had a week to go before we had to go home and I was no closer to determining the problem with the equipment when it was in flight. We had checked all the aircraft electrical but still could find nothing to indicate an intermittent fault. When I found the solution some 6 months later I felt an absolute ass – hindsight is always 20-20.

******

When it was time to return home we were picked up from Normanton Airport by Bush Pilots for our flight back to Cairns to connect with the flight to Melbourne. Bush Pilots run a service that is more like a bus line than an airline. They drop into station airstrips and drop off supplies and/or pick up passengers and any concept of a time table is coincidental. Accordingly by the time we were on descent into Cairn we had a nice view of the QANTAS flight to Melbourne just taking off. "Oh tsk," he thought another night in the tropical north. We went over to the QANTAS desk and made a booking for the next day and then realised we had no place to stay. John suggested we go back to the guest house and see if they could help but in the days prior to mobile phones and credit cards it meant a cab ride and a cheque to pay the rent.

We were nearly out of cash but I had my cheque book so off we went. They had a room and we dumped our gear and walked into Cairns. We went to the Marlin Bar and were having a drink whence got into conversation with a trawler skipper and ended up drinking with him for the rest of the night. He then paid for the dinner we had and we left him late at night in good spirits and ready for the flight home to the cooler climes of Melbourne.

****

Home again I was given two interesting pieces of news. There was a large passionfruit vine on the side and back fences and Carol was getting nauseous when she smelt the fallen fruit.

She was pregnant again and it was easily traced back to the one night I was home between the equipment course in Sydney and the Tac1 course in Canungra.

****

Chapter Three- - Geoceiver ME in USA

That was exciting home front news but Eric Lauritzen had equally exciting work-related news for me when I went in on the Monday. RASvy was procuring from the United States a satellite geodesy system called The AN/PRR 14 – Geoceiver and he had chosen me to go to the US and conduct a Maintenance Evaluation. John Strautins would be coming with me once again.

The Geoceiver used the Doppler shift of signals sent from the Transit satellites to determine positions on earth. It was designed to be a man-portable satellite tracking station consisting of four parts: antenna, receiver, amplifier, and punched tape recorder, fitted into specially constructed stiff-foam padded carry cases, and was suitable for the accurate mapping of relatively inaccessible or remote areas

A Maintenance Evaluation (ME) involves the detailed understanding of how the equipment works and how to repair and maintain it when it is deployed in a military setting. Consideration must be given to:

- The conditions in which the equipment is to operate,
- The skill level of the operators and technicians using and maintaining it,
- The training necessary to maintain and/or repair the equipment, and
- The tools needed to do the support task.

Therefore, the more important aspects of any ME include the identification of the need for specific-to-type repair and maintenance, the gaining of the knowledge required to perform these tasks, the identification of specific test and repair equipment needed to conduct this necessary function, the provision of satisfactory repair documentation to facilitate this, and the ensuring of a ready supply of repair parts for all levels of repair.

It is obviously not sensible to purchase equipment that we cannot operate or maintain and repair. RA Svy had the operator problem and Raeme, i.e. John and I had the maintenance and repair problem.

Background. In October 1957 two American physicists observed the Doppler shift of radio signals from Sputnik, and realized that they could use these signals to determine the orbit of the satellite. This was extrapolated later to allow the use of Doppler signals from satellites in stable orbits to produce an accurate, passive, all-weather, and world-wide navigation system, NAVSAT particularly for the Polaris Submarines.

The Applied Physics Laboratory of the Johns Hopkins University, the prime scientific contractor for the Polaris program, developed the Transit satellite system with a constellation of five required to be operational at all times to ensure reasonable global coverage . For redundancy purposes, ten satellites were usually in orbit. Unfortunately with the orbits crossing the poles and spread out along the equator you had to have a calendar of orbital crossings, quoted on what is called a Julian calendar, to ensure you were ready when a satellite was above the horizon and visible to the receiver antenna. The delay at the equator was several hours and about an hour or so at the tropics.

In operation the satellites transmitted two UHF carrier signals (150 MHz and 400 MHz) providing precise time updates, within 50 microsecond accuracy, every two minutes and other ephemeris data giving the satellites six orbital elements and orbit perturbation variables. This ephemeris data was essential as the gravitational effect of travel in orbit over the changing earth profile and the rotation of the earth itself caused incremental changes to be observed by the computer aboard the satellite and returned as phase modulated data to the ground station. The ground receiver computer used this to adjust the orbital calculation. Two carriers were used to compensate for ionospheric refraction. The navigation/surveying software used a 'least squares' best fit for each two minute portion of the Doppler curve, thus the more passes recorded from a stationary ground station the more precise the result. For surveying purposes it was usual to record orbits over a minimum of twenty-four hour, and usually a week so as to archive accuracies of plus or minus 5 cm.

The Navy released Transit signals for public use in July 1967 and started the movement into satellite navigation in pleasure and racing yachts. See more later.

John and I spent time with the RA Svy project people and it was determined that the purchase of the equipment would be largely dependent on our satisfying the maintenance and repair question. This equipment was expected to be full military spec and the repair and maintenance provided had to meet that requirement also.

To enable us to achieve our aim and prepare the detailed ME we had to read as much as we could about the principles of satellite geodesy a completely new field world-wide let alone in Australia; all this in pre internet days. RA Svy had some papers and contacts in the US, particularly at The Defence Mapping Authority (DMA) near Washington DC and we read these but they tended to be more operator then maintenance oriented so it was going to be a case of sitting down with the engineers and technicians if we were to gain the information we needed.

Funding for the ME was made available for expenditure in FY 72/73 so with RA Svy keen to have an early result it was agreed that we would fly to the USA on 10

November 1972 and spend roughly five weeks visiting the prime contractor Magnavox in Torrance, just outside Los Angles, the US Army user, DMA, near Fort Church Virginia, and the US Air Force user, the 498 Geodetic Svy Sqn at Francis E Warren Air Force Base (FEW) just outside Cheyenne Wyoming. Our in-country sponsor was the Australian Army Staff in Washington DC, AAS (W) so we had to make time to visit them and establish contact in case of the need to address admin details.

From a technical viewpoint we were an autonomous entity and could operate unsupervised. They, AAS (W), would have a copy of our itinerary but with communications as they were then, it was a default system only. If something screwed up or was of real importance we or they would make contact. Considering Carol's pregnancy there was, obviously, one piece of important information that I would be most anxious to hear. The doctor basing his prediction on general information of period cycles etc. and ignoring the reality of 26 February 1972 had predicted the baby would be born about 15 Nov 1972 we *knowing better* calculated 29 Nov 1972. Whatever the date quoted it was still to be seen if Carol could carry to full term. After Andrew the medical opinion was not positive but we lived in hope. In either case if Carol did go full term I would be in the US so I set up a means by which I could be contacted as soon as possible. My boss in Albert Park would send a signal to AAS (W) and they would let me know all the news

Given the important bit was out of the way it was incumbent on me to arrange the visit details and ask the travel people to make the necessary bookings and calculate our travel allowances(TA). Because we were going to be in commercial accommodation, hotels or motels, and buying our meals we were on full TA.

Finally it was decided we were going to fly out of Melbourne and land at Los Angeles (LAX) in the evening of

[Type text]

Friday, 10 November 1972 and would be picked up from our
accommodation on the Monday morning by Magnavox and
looked after by them before making a courtesy visit to AAS
(W), during a week with DMA talking repair and
maintenance with the Army. Our next call would be with the
USAF at FEW near Cheyenne for another period gaining
more hands-on experience testing and repairing the US Air
Force equipment. Our final few days would be back at
MAGNAVOX discussing our findings and clarifying any
doubts or discrepancies. We would fly out of LAX to arrive
back in Melbourne on 19 December 1972.

Now that we had our itinerary agreed it was important
we deal with a couple of minor matters, the first of these was
to get passports with visas to cover us and that was a job for
the AHQ staff and all we had to do was arrange for photos
and they did the rest. Seeing that I had never bothered to
replace the watch stolen from me five years earlier it was an
opportune time to do so duty free. I went up town in
Melbourne to a duty free shop in Swanston Street and found
a Seiko auto-wind divers' watch for $25 and snapped it up. I
still have it and have been offered $300 for it. Because we
were going to harsher climates than usual we were allowed to
buy an overcoat at the Queen's Expense, so I bought what I
thought was suitable a tan coat with a zip out wool lining.

We flew out of Tullamarine via Sydney on a BOAC
707, landed at Pago Pago, then Honolulu where we cleared
US immigration then on to San Francisco, arriving late
afternoon. As mentioned before all flights were first class as
we were spoiled by time we arrived. After a short delay we
caught a domestic flight to LAX landing as the sun went
down. We passed through customs showing only our carry
on board luggage because our bags were still in San
Francisco. Luckily we had been given a Qantas carry-on
shoulder bag each and we both had our toiletries in these so
were alright for the moment. Qantas assured us the bags were
following us on the next flight and would be delivered to our

22

hotel. Next door to LAX is the Airport Ramada and this was where we were staying over the weekend. Resigned to this problem with our luggage we took a cab and booked in and went to our rooms after making arrangements to meet in the bar for a drink and a meal. The bar was in fact a piano lounge and Lennie Blewitt, an African –American or black or coloured or whatever, piano playing singer and was very good.

This was not that long after the Watts Riots in1965 and so were unsure what was acceptable terminology so decided to call everyone 'Mate ' or 'Lady' and avoid any race or colour. For example when telling others about something an African-American did the generic he or she was used.

We had a bite to eat and then went and sat at the bar near the piano and ordered some drinks. We started with beers but after tasting a couple of different types went to Johnny Walkers Black with a touch of water, and stayed there. There were a few people in the bar when we arrived and our accents soon had the question we were to become used to. "You have a funny accent. Where are you from?" Everyone who asked was polite and simply curious and it became an easy ice-breaker for the rest of our visit.

We had been given our TA in $US travellers' cheques and decided to have our breakfast and main meal in the hotel and charge it to the room. For our drinks it was easier to run a tab and pay with a travellers' cheque meaning we got our change in cash. So on the first night this was the plan. John was at the bar talking to a young girl and I was sitting at the piano talking to Lennie and it was a good night but eventually I folded and left for an early night, arranging to meet John after breakfast.

John, when asked, said the girl, Linda had gone home just after I left, apparently she lived locally, and he had only stayed for a short time then went to his room as well.

23

The next morning when we awoke our bags were in the hotel as promised. We took a taxi to Pueblo de Los Angles the original site of the settlement of Los Angles then an open air Mexican-flavoured market. We arrived about lunch time, so found a small bar that sold food, Mexican food. We had no idea what some of these were so punted for burritos.

www.taste.com.au/recipes/25550/beef+and+bean+**burritos**

Unfortunately the burritos we had were made by taking some minced beef, tossing it in a pan and emptying the contents of their vast array of spices on top then frying and refrying it all, until it was nearly capable of spontaneous combustion , wrapped it in a tortilla and passed it to the unsuspecting Anglos. They sold plenty of beer that way.

It was a lovely day and we roamed the streets and piazzas admiring the old buildings and checking out the stalls offering Mexico at a discount. Sitting in the bright sunlight our attention was drawn to a group of young children. One at a time, they would be blindfolded and swing a long stick trying to hit a bull shaped Chinese paper lantern suspended from a post about five to six foot above the ground. It was a kid's birthday party and the Chinese lantern was a papier mache piñata full of lollies, chocolates and little toys. When they managed to split it the cascade of goodies were spread among the cheering kids. It was a fun time and the love of all those involved was obvious as all attempts were cheered and applauded. We left with a smile on our faces and caught a cab back to the Ramada.

Agreeing to meet later we split up to have a shower and change before eating and gravitating to the piano bar where Lennie welcomed us with "Tie my Kangaroo down Sport". It was taken up by half the crowd so I think we may have been set up. I bought him a drink and we sat down for a good quiet night. The girl from last night was there again and John went over and sat with her as I talked to Lennie again, this time about American Football which was a subject I knew nothing about but he was a walking book of statistics. He was one of many Americans who followed the game very closely and could tell you how many passes a quarterback threw and how many were intercepted or any other remote piece of obtuse information you did not need to know. He explained the difference between pro and college ball and when you could see them. Monday night apparently was the big night so I promised I would watch on Monday. During this mainly one way conversation he mentioned that Linda had asked for me when she arrived. I looked over at John but Lennie said, "No, you my man. She asked about you."

Seeing me glance over, Linda picked up her drink and came over and sat next to me at the piano and we started talking. John joined us for a few drinks but Linda was making it plain that she wanted to spend the time with me and he left. It was now about 2200 hrs and we continued for another hour whereupon she asked me to spend the next day with her. "To keep me company." was how she described it. Lennie was finished for the night and smiled and waved. Like an idiot I was not thinking just pleased with this surprising and, I'll admit flattering, proposition and I told John that I would see him late Sunday.

The next day Linda picked me up and we went to Disneyland and I spent a day being a kid again. I know I was doing something stupid but that selfish element of my personality had taken over and it was me thinking only of me. The only unselfish thought I had was a promise to bring the kids to Disneyland one day. It took four years but I managed it.

****

Linda dropped me back at the Ramada afterwards and I packed up my gear ready to be picked up at 0800 hrs by the Magnavox rep who would take us to the factory in Torrance.

Torrance is in the SW region of **Los Angeles County and** has 2.4 km of beaches on the Pacific Ocean. It has a moderate year-round climate with warm temperatures, sea breezes, and low humidity to me it was not unlike WA, except it was facing a different ocean.

First stop was for breakfast then we went to the Hi-Ho Motel near Redondo Beach and close to the Magnavox factory. This was to be our accommodation whilst at Torrance for a week from Monday 12 November 1972 until Friday 17 November 1972. Our days followed a set agenda: Breakfast with a Magnavox engineer, drive to the factory and spend the morning learning how to operate and maintain the equipment, followed by lunch with a couple of engineers then an afternoon back on the test beds. the chief engineer was named Bob and was an ex US Navy engineer, in fact all of the team at were ex-services and had much experience in navigation , rocketry , aeronautical engineering or the like.

Magnavox as well as being the prime contractor was also the holder of the maintenance contracts for support of all the Geoceiver based geodesy or navigation systems deployed around the world. Most were in fact navigation systems aboard tankers or container ships and the engineers of a rotation basis were on immediate call to up tools and fly to wherever the vessels with faulty equipment may be. This type of experience was of great value to us and we pumped them for all the information we could. After work they would take us out for happy hour then buy our dinner at a local restaurant before dropping us back at the motel.

I found that the US meals were different to what we were used to. All meals started with a glass of water generally iced water, and were followed by a salad. Salad over there meant a serving of lettuce with an accompanying dressing. A salad as we knew it may have been called a garden salad but we decided 'when in Rome etc.' and did not query it. One other marvellous discovery was real meat well prepared. In Australia we did not know how to age beef and I became addicted to Prime Cut steak and had it most meals.

It was so tender it seemed to be able to be eaten with a fork and tasted superb. **Speaking of eating with a fork we did not change our style of using** knives and fork. It would have seemed too pretentious.

In all of this time we could buy an occasional drink but not often. We were there guests and they treated us royally. By the end of the week we were able to set up and operate the equipment, conduct operator testing to ensure it was functioning properly, using the built in testing systems, and finally take it apart down to the base chassis and rebuild it and test it. The most significant module of the Geoceiver was its crystal oscillator which had to remain accurate to within very fine margins if the phase modulated signals and the Doppler shift that was measured would provide meaningful data. This was the principal test we did each time we fired up the equipment.

Given that we had to work to the window of opportunity granted to us from the time the Transit satellite rose over the horizon and before it set again we had to keep the oscillator running for long enough and early enough to enable it to become stable, which ideally meant having power on to the equipment full time whilst deployed. During field deployment that critical requirement necessitated us having a stable power supply on hand at all times to keep the oscillator happy. This was to be very important to me later on.

****

The Saturday John and I were taken by one of the Magnavox guys to San Diego Water world where we played tourist and on Sunday morning we were off to Washington

[Type text]

DC. Our flight to DC was with American Airlines and again first class. I had my Qantas carry-on bag with me, all maroon with a flying kangaroo and Qantas all over it, and we had not been going long before I was approached by a stewardess who asked if I wanted to go up to meet the pilot. It turned out they had me on the manifest as Captain Mason and because I had my Qantas bag assumed I was a Qantas captain. I corrected the error before they asked me to fly the rest of the way to DC.

On arrival we were met by an embassy car and taken to The Embassy Hotel, on 2015 Massachusetts Avenue, Washington DC, and registered. The room had been pre-booked. The next day, Monday, we had to report to the Australian Army Staff Washington (AAS W) at 0900 hrs to sort out the administrative details of our visit, and arrange to start work at DMA.

Just to fill in some time that night we wandered over the road to Flanagan's Bar and tried Carlings beer. Not bad. Considering our need to be bright-eyed and bushy-tailed for our next morning appointment we did not tarry, but before we left we noted a sign that promised all the beer you could drink for , I think, $10, could have been less(??) "Might check that out if we had a chance." we promised each other.

The next morning Monday, 19 November 1972 we were up at 0700 hrs, too early for breakfast which was not available until 0730 hrs in the dining room. No problem we will go for a walk and get familiar with our new temporary home. Accordingly we went out and turned right and strolled up a way and decided to go right for a couple of blocks then complete a square back to Massachusetts Avenue and our hotel, in time for the dining room to open. Down Q Street across some others and back up P Street. Bit of a rundown area seeing we were so close to everything but shrug! Back in time for breakfast and after trying to work out what the various types of eggs meant, scrambled meant cracking the egg on a hot plate and slashing at it with a spatula or knife cutting it into ribbons (?) we opted for a Spanish omelette. I had had one in LA so knew what it was.

At 0900 hrs on the dot we fronted the RAE Colonel who was the Australian Army Rep, Washington and had our first surprise. Just about the first thing he said to us was to be very careful at night in the city as it was quite dangerous on the streets. Particularly do not go past P Street towards Q Street, R Street etc. as that was a no-go area. Now he tells us!

All was fine on the financial side of things and we were able to acquit the TA we had spent and claim back the odd other expense not covered by TA. We were given a voucher to go to the nearby AVIS Hire Car depot and pick up what was described on the voucher as a mid-size sedan. With this, and a map to show us how to get to the Falls Church Motel where we would be staying, we would soon be off again into the wilds of VA John took off to pick up the car and brought it back to the embassy. With a car to get around in, and a new city to drive in we did not loiter and having checked out of The Embassy Hotel in the morning was soon on our way down the Custis Memorial Parkway (Interstate 66), past Arlington Cemetery and on towards Falls Church and the Falls Church Motel another temporary home.

The mid-size sedan turned out to be a Plymouth Fury and you could be mistaken for thinking it was an aircraft carrier, it was so large with its flat boot and wings.

The colonel invited us to his house for dinner and to watch the Monday football, he had become a fan and never missed a TV game even college ball. It was a good night and he asked us to join him and his family for thanksgiving on Thursday night.

****

During the rest of the week we made visits to the DMA HQ at Springfield Va. and talked to the US army equivalent of RA Svy. They explained to us their methods of deployment and how they intended to support their operations. They had set up a facility in nearby Falls Church which provided what we would have called field repair. Given their equipment numbers they could replace one-for-one defective equipment and complete module replacement repair at Falls Church then return the defective modules to Magnavox for repair. The Army repair facility was set in a wooded area lit with the first flush of gold of the northern autumn. A two storey wooden structure staffed with mainly civilian or ex-military tradesman it looked more like a bushland retreat than an army workshop.

We were taken in hand by John, a huge African-American (Just as an aside, American men were often much larger than Australians.) who was one of the biggest men I had ever met. I will call him US John to differentiate him from our John.

US John was completing a diploma of electronics and while we were there he told us he was preparing to do his final test. His knowledge of the Geoceiver equipment was total and he gave us some insights into testing that he had developed and was only too happy to pass it on to us.

At the same time as we were visiting there was a team from the UK army doing their own trial deployments along the USA/Canadian border and they had determined errors in the agreed border of up to 50 meters in places.

Of more interest to us was that they were operating out of a Land rover, Fitted for Radio (FFR) with a 24 v system that they were using to operate the Geoceiver, through a separate stabilised power supply. Like everyone else they had had some difficulty when they had to relocate the equipment with maintaining the oscillator stability. When they were using the Land rover on-site it was fine but if there was a need to move without the resources of the vehicle like in a helicopter or aircraft it was problem

We sat down and examined what they were trying to achieve and came up with a solution. Given that the four components of the Geoceiver came in very thickly padded carrying cases that would provide protection from both shock and heat/cold it was considered prudent to use these as part of our solution.

The stabilised power supply was placed in a recess cut deep in the lid padding with enough padding above and below the inserted power supply to ensure it was still protected. We only had to run the inputs to the side of the case and the output to the receiver and we had an integrated carry case and power supply that could operate from any 12 /24 v supply. We drew up a block diagram and took photos for future reference.

The UK Team were going on another deployment to Montana to evaluate the equipment in extreme cold. It was arranged for us to meet up with them later at the USAF repair facility. At that time we could see how successful the modification had been.

Tuesday night we decided to check out Flanagan's and sat down and said we wanted the all-the-beer you can drink

[Type text]

special. Out came a jug, a 'pitcher' of Carlings Black Label
each, and we sat down ready for a session. Having emptied
our pitchers we went for a refill to be told the all you can
drink meant all Americans can drink a pitcher. We explained
the reality of false advertising and assured him that a pitcher
was just a warming-up drink to Australians. We all laughed
about it and he kept the pitchers coming as we chatted with
him throughout the evening. Doubt they made any profit that
night.

On the Wednesday night we were invited to a party at
one of the engineers. It was intended to celebrate US John's
completion of his diploma. He had sat his final on the
Tuesday and got a mark of 98%, outstanding. Luckily we had
been able to stock up with Johnny Walker black Label at the
embassy so we had our entrée card. US John was in fine
form and it turned out he had a fine singing voice and played
the guitar as well. 'A MAN AMONG MEN!'

The Thanksgiving Dinner on Thursday 23 November
1972 was another good night and we met the AAS Rep's
aide, and his family and they were after us to update them
about the upcoming Federal Election, due on 2 December
1972.

This reminded us that we also had to vote so it was a
case of two birds with one stone as we were also invited to
the end-of-month Happy Hour that Friday. On the last Friday
of each month the embassy staff and guests meet down in the
basement to eat seafood and drink Australian beer. As the
comedians say," Timing is everything."

It was imperative that we amend our itinerary so Friday
before lunch we were back at AAS W, and voted, before
sitting down with the AAS staff to make changes to our
itinerary. As a result of what we had observed at DMA, it
was agreed we would return to Magnavox for a week and go

to FEW a week later. With a new itinerary and ticketing finalised we attended the happy hour.

****

Our weekend was one of tourism. First we went to the Arlington War Cemetery and spent an interesting time among the rows and rows of the fallen of US military in successive wars. One part of the cemetery of particular was of special significance.

It was the grave of President John F. Kennedy. The grave is marked with an 'eternal flame'. The remains of his brother, Senator Robert F. Kennedy is buried nearby. The latter grave is marked with a simple cross and footstone.

In OCS we had studied military history and the campaign we studied was the Shenandoah Valley campaign of the American Civil War. It was a favourite subject as far as I was concerned and I was delighted to find myself at the northern ends of the Shenandoah Valley with a day to explore and a car to do it in. John was not interested so I took off on my own very early and drove along the route suggested in a brochure from the hotel. My first stay was in Manassas where on July 21, 1861, the two armies, the Union and the Confederates clashed for the first time on the fields overlooking Bull Run. Heavy fighting swept away any notion of a quick war. In August 1862, Union and Confederate armies converged for a second time on the plains of Manassas. The Confederates won a solid victory bringing them to the height of their power. It was only about this time that I realised how far north the Confederate states were and still are. The confederate flag is seen everywhere just out of Washington DC.

Following my brochure I went south and visited the so-called Field of Lost Shoes. The Battle of New Market was fought on May 15, 1864 where a makeshift Confederate army of 4100 men, which included 247 cadets from the

Virginia Military Institute (VMI), had forced Union Major General Franz Sigel and his army out of the Shenandoah Valley. During the battle the cadets, as part of a confederate push towards the union lines had to cross an orchard that was so muddy that some lost their shoes. This battle has been the subject of two feature films, one I had seen as a younger man and enjoyed, the latter film was released in 2014 but has not been a huge success. The field of lost shoes

The Americans do their history very well and there is a brilliant museum with extremely informative dioramas and exhibits and I spent a wonderful time following a tour group listening in and learning more in an a few hours about the reality of the civil war than in a semester of study.

There were 247 VMI cadets involved in the battle of whom 10 were killed and 47 injured. During the battle five Union guns (artillery pieces) were captured, and the cadets captured one of them. To put that in context the capture of a gun is the equivalent of the capture of a regiment's battle flag, and in this battle when Sergeant James Burns of the 1st West Virginia Infantry saved the regimental flag in the battle he received the Medal of Honour.

With my tourism over I returned to Falls Church and readied myself for our next little adventure

During our alone-time we wrote up that week's activities ready for the weekend where we would pack up again and prepared ourselves for the next week.

This Monday 27 November 1972, our week started with another cross country flight from Washington DC to Los Angeles. Our flight was just after lunch and we would take off and arrive about mid-afternoon given the three hour time difference between the East Coast and the West Coast

Our departure from the east coast was through Dulles Airport and here we were stuck by another of the minor irritants of life in the US. When we settled into the airport lounge awaiting our flight John went to the toilet and I asked the waitress for two Carlings. She told me I could only order one drink at a time. I said," It's alright the other one is for my friend who had to go to the toilet".

"No problem," she said, and put my glass in front of me on my table and the other, John's, on the vacant table alongside. If this type of stuff was country wide we could have adapted but it changed state-by-state. Some were totally dry but just over the border you could drink anything you liked. It was a problem for those of us who liked a drink on a warmish day and was travelling all over the country. Nowadays there is probably a website to help you fathom the variations but then it was hit and miss or, really, trial and error.

Bob, the Magnavox chief engineer and now a real friend picked us up and took us back to the Hi Ho motel and as it was near Happy hour we celebrated our 'return home.' and he bought us dinner. They Magnavox were simply marvellous and we felt a part of the company.

The next day 28 November 1972, was spent in discussion with the engineers what we had learnt and they were particularly interested in the idea of building the power supply into the Case padding and we showed them the diagrams and photos we had taken. They could see the plus side of it for on-ground deployment in remote regions, which of course was our raison d'etre when we were deployed as a military survey unit and its support element.

We had, nearly, the run of the workshops except for one security area that was off limits. Do not know what was behind those locked doors and did not ask. I was in their environmental testing area, when I saw something very familiar it was a PDP 8E computer and it was sitting on a vibrating table. The memory board was exposed and you could see what was happening to the ferrite cores. They were vibrating and approaching harmonious standing waves as the period of the vibration was adjusted. It did not need a Faraday to see that the movement of the copper wire supporting frame work through the magnetic fields of the ferrite cores would generate an Electromagnetic Force (EMF). These randomly generated EMFs would scramble the data stored. What would the vibration of an aircraft do to the memory in, for example, Aerodist? I contacted RA Svy immediately and suggested they look at their problem in light of this information.

****

Sometime during the late afternoon of Thursday 30 November 1972 I received the message from AAS (W) that had been expected. David Patrick Mason was born the day before i.e. 29 November 1972 at 0430 hrs Los Angles time or 2130 hrs in Melbourne. (8lb 9oz) The message said both well. This was later proven to be far from the reality.

Carol had been in labour for the best part of a day and David was delivered using forceps leaving his skull misshapen and Carol requiring stitches and in lots of pain. If it was today it would have been a caesarean birth and both would have been well. That doctor (??) was lucky I not there or things might have gone badly for him. Carol's GP, her doctor since she was a baby, was appalled when he found out but it was too late then. Given her prior history with pregnancies one can only wonder what was going through this clown's mind. Carol suffered postpartum depression and was not well for some time. I did not make it home until David was nearly three weeks old and Carol had her family with her, but no husband.

I immediately told everyone we had been working with, and a celebration dinner was planned for the following night. One of the engineers, the Texan asked, "Have you spoken to Carol yet?' I looked at him in bewilderment and reminded him she was in Melbourne, Australia. "Not a drama my man," he responded" give me her number." I did not have it but told him she was in Frankston Hospital, Frankston, Victoria. Given the fact that Melbourne was 17 hours behind LA we had to do some calculating to determine when both cities were awake.

We worked it out that if we rang at 0800 hrs LA time it would be 1500 hrs the day before in Melbourne so he picked us up early and we had our breakfast and was by his desk just after 0800 hrs.

This was my first experience with direct dialling. He picked up the phone got a line dialled the code for Australia, then Victoria and finally the number and I was talking to the switchboard at Frankston hospital. It sounds all so easy now but appeared to be all done with mirrors and piano wire back in 1972. After being put through to the nurse's station in Maternity, I had to wait while they dressed Carol, put her in a wheel chair and wheeled her to the phone, as she could not walk on her own. We had a good long chat on Magnavox's money and then she had to go back and lie down.

The celebration dinner was held at a Japanese restaurant in Century City, bordering on Beverley Hills. We all sat around a low table with our feet in a well, to give an impression of kneeling. The menu was of a banquet type and I felt obliged to sample as many as I could, mixed with a selection of warmed sake. With a background of soft Japanese style music the atmosphere was one of peace and tranquillity. The method and location of the wetting of David's head was one I could never have possibly imagined, just months earlier.

****

On the Sunday we were to fly to FEW and begin our period with the USAF. We arrived at the Denver airport in the middle of a snow storm and hurried into the terminal across an open tarmac. Denver is called the mile high city because it is just on 5280m ft above sea level and is in the rocky Mountains, we stayed only for a short time, about two warming JW Black Labels, then we left for the 100 mile flight to the even higher Cheyenne, 6100 ft above sea level.

A young aircraftsman named Jim met us at the airport and said he was our driver for the total time we were in Cheyenne. He took us to our hotel and left us there, promising to pick us up at 0900 hrs to take us to FAW. We decided to test the Wyoming liquor laws and had dinner in the hotel restaurant and then a few drinks in the bar. I ended up talking to the press secretary for the City of Cheyenne and he was an amusing and enthusiastic promoter of his city.

I learnt that just outside Cheyenne was the largest truck stop in the entire USA, Little America, plus a few other 'largests' that I cannot remember now, 'rock crusher'??. He promised to take John and me to this magnificent truck stop the following night and with that and a need for sleep we left him.

Meanwhile back in Australia the. Federal elections for the House of Representatives were held on 2 December 1972, and won by the Australian Labor Party led by Gough Whitlam. The elections ended 23 years of successive Coalition governments which held power since 1949. The US yawned. We finally found the result, 3 column inches, on page 26 of an LA newspaper. I was to find a more detailed coverage later but still buried deep behind 'The real news'. We learnt that Gough had taken unto himself all cabinet positions sharing the glory only with his deputy Lance Barnard. A cartoon in the Guardian Newspaper I saw a fortnight later showed how others looked at this situation. It showed a big Gough dressed like a clownish one-man-band with policy statements falling from every pocket and behind him a tiny Lance Barnard trying to pick them up and make sense of it all.

Jim turned up in the morning as promised and we went to meet the general who was CO of FEW. We then were taken on a tour of the large base. It is in fact a missile base with a few aircraft squadrons on site but mainly it is to support the surrounding Minuteman missiles in their in-ground silos.

In 1972 they were a very important element in the cold-war. Unfortunately the pressure of the trip had caught up with me and I had to beg off the tour of the Minuteman silos in the afternoon and returned to the hotel for a sleep. John woke me when he returned and we went down to meet the advocate of Cheyenne and have dinner at Little America.

The 498 Geodetic Svy Sqn at Francis E Warren Air Force Base (FEW) operated out of centrally heated hangars and each morning the Geoceiver on site were run up and operator testing was conducted. This amounted to setting the device to self-test and scanning the paper printout that was punched out on the punched tape recorder. The output logic was reverse polar where high on the output generated a Zero, and a lower or no voltage generated a one. it was an eight digit read out and if everything was fine you would have a readout of all zeroes . Things were going along fine until one day we came in and they had one machine pulled apart and were rapidly exchanging modules and doing self-test after self-test. When we asked them what the problem was we were told, "All zeroes when left on the bench, but something is definitely wrong, because when we try to take it away and return it to the case it seems that the movement causes faults to occur. We have to keep changing modules until we find which one is causing the fault."

John and I looked at each other and went for a coffee and when we came back offered to have a look later if they had no problems with that. By this time the techs who had been working on it were so frustrated that they were happy to hand it over. To be fair I should add that the Master Sgt who was their boss was not present that day.

Like the USAF techs we took it out of the case and ran the self-test, sure enough all zeroes. Next we put it back in the case and tested it again - all ones. This was just what the techs had described but they had missed the clue that the printout was giving them. If the readout was all ones it meant no voltage was getting to the display logic. We carefully reassembled it into the case, taking care that the eight wire ribbon cables were clear of the case and not shorting out. A quick run of the self-test and the tape showed all zeroes. We went to lunch and discussed whether we should tell them what the problem was and make them feel stupid or simply leave without explanation and save them embarrassment. We decide to not explain and let them think we just got lucky.

Later we explained it to the Master Sgt so he could deal with it how he liked. I assumed he replaced the entire wiring loom but do not know for sure.

●WEATHER

First figures indicate highest temperature during the past 24 hours; second, lowest temperature last night; third, rain or melted snow during the past 24 hours, ending 7 p.m.

Anchorage	23	18	.00
Big Piney	4	-21	.03
Billings	-11	-20	.00
Boston	62	29	.47
Casper	20	-24	.00
CHEYENNE	6	-24	.00
Chicago	16	12	.20
Denver	3	-17	.00
Dodge City	11	-6	.00
Douglas	5	-31	.00
Evanston	19	-8	.02
Ft. Worth	33	21	.00
Grand Island	7	-7	.00
Honolulu	80	63	.00
Kansas City	11	0	.02
Lander	-14	-26	.00
Laramie	19	-33	.00
Las Vegas	49	25	.00

Cheyenne is high in the Rockies as I have said and when I decide to go into town for a haircut the temperature displayed on the screen outside the bank said it was just 2 degrees Fahrenheit (-17 degrees Celsius)and this was at midday. The temperature that night fell to minus 24 degrees Fahrenheit (-31 degrees Celsius) and the maximum was 6 degrees Fahrenheit (-14 Degrees Celsius). See cutting from The Wyoming Eagle dated Thursday 7 December 1972.the lead story that day was,' THE LAUNCH OF THE APOLLO 17 MISSION AFTER THREE HOUR DELAY'

[Type text]

Now without hair and without a hat it was time to get indoors and I went to the Cheyenne Museum. it was a fascinating place outlining the lives of the plains Indians, Cheyenne Arapaho, Crow and Sioux . Having seen the modern bows it was intriguing to see that their bows were only half the size of a standard bow today.

Our work at FEW continued until the end of the week and we sat around over the weekend writing up our findings and with Jim still available as our driver we went out to a nearby National Park area and saw a small herd of Bison in their natural environment. I never realised how large they were, and when we read the information sheets available, how many there had been before the' Buffalo Bills 'of the 19th century nearly wiped them out.

The rest of our time at FEW we continued our testing and dismantling and rebuilding the receiver module until we could do it blindfolded. We were in the hangar when the UK team arrived back, and spent a day or so debriefing them and received a promising report on the modifications they tested. They had been in even worse weather in Montana than we had in Cheyenne and it was good to hear that everything worked well and there was no problem with the robustness of the packing as result of our implanting of the power supply into the foam padding.

Given that LA was a trifle warmer than Cheyenne ,we would only be repeating tests we had already done many times, and we had exhausted our debriefing with the Brits, We decided to return to Magnavox on Tuesday, a few days before we flew out on Sunday 17 December to go home. The purpose of this last visit was to discuss on-going repair and maintenance support between the Australian Army as the user and Magnavox as the designated repair contractor. It was part of my job to build a workable repair system and I had to establish that willingness to participate and the capacity of the involved parties. The actual negotiations and

42

final contracts would be followed up by the supply people in AAS W, acting on my recommendations.

****

Finally on Sunday 17 December 1972 we flew out of LAX on a BOAC flight and it was on board this flight that I saw the cartoon lampooning Gough.

We lost a day coming home and landed at Mascot on the morning of 19 December and waiting in the transit area, not having cleared customs or immigration. That was to be done in Melbourne. Then we spotted a kiosk selling beer, Australian beer, "Why not? "We said, and asked for two cans, "Toohey's or Reschs" she screeched at us. "God Almighty" we thought "We are home!" We had become so used to the softer vowels of the Americans that our own accent grated harshly. Later on, by the time we had reached Melbourne, we realised it was just that poor woman.

Prior to getting back to the harsh reality of Australian Army life we went on leave and I was introduced to our new born son David Patrick. Carol was still not completely over the birth of David and he was still showing the forceps marks but they were beginning to fade. It was surreal leaving with two children a girl and a boy and retuning to find another child, a boy ha d crept into our little circle. I having failed to bond with him at his birth and having missed being the care giver and supporter at this crucial time felt distant and remote and it took some time to come to accept and learn to cherish him. Luckily we both grew out of it and I can honestly say that now I admire the man and particularly the father he has become.

After some leave, we mainly me, set about the writing up of the ME with John providing the valuable technical hands-on view of the repair philosophy. During our time with the various agencies I had been drafting what would end up as our ME report and it was simply a matter of massaging these drafts into a coherent whole, detailing the support programme that we would use to maintain and repair the AN PRR 14 Geoceiver.

Following the lead of the US army we recommended RA Svy buy redundant equipment and replace one-for-one, i.e. Operational equipment for defective. Raeme would fault find to module level and the defective modules would be returned to Magnavox and they would do a one-for-one swap. (New for old at 40% of new price.)

That is they would send us by express delivery a new or refurbished module and we would send them back our faulty one so they could refurbish it and place it in stock. The cost to us of this exchange program was set at a rate 40% of the new price.

When writing up the ME report I wrote an extract giving one of, or conceivably, the first paper seen at WAIT on Satellite Geodesy and presenting a diagram and photos of the modifications and enhancement of the power supply as a Project Design paper. It was accepted and I was awarded my Associateship and attended the Graduation ceremony at Bentley Campus in April 1973.

Chapter Four – Operation Gading 3

While we had been away, the Directorate of Operations were planning the operational programme for 1973 and there was a meeting at the survey regiment in BENDIGO very soon. Two of us had to attend the planning meetings as the Raeme reps and advise on the Raeme support available, Maj Bob Millar who looked after the Army aircraft side of it and me the electronics aspects.

It was necessary to establish a core group of technicians to support Survey and I had to extract these techs from the current technician radar c course (Tech radar) and cross train them onto the survey equipment, tellurometers, Aerodist and introduce them to Geoceiver. We created a new employment category of Technician Electronic, Ground (TEG) and to allow me to train and test the TEGs I was made a Trade Testing Officer, ECN 224, Tech Radar. Over the next month or so we trained and qualified six new TEGs.

The next RA Svy operation was operation GADING 3, and the unit involved was 5 Fd Svy from the artillery barracks in Fremantle, yes that place again. The Aim of the Operation was to complete a survey of the northern end of Sumatra in Indonesia and establish a geodetic connection a reference between Malaysia and Indonesia over the Johore Strait. The operation was to be based at Polonia International airport just out of the city of Medan in Sumatra. Our six new TEGs were to be deployed in support. When bids were requested for travel I had made one for a short visit to Gading 3 to assist/observe as relevant, RA Svy agreed.

Although the Aerodist was to be deployed it would act as a recording system, not the complete processing system that it was supposed to be.

45

An integral part of the Aerodist system was a Hughes 7 track magnetic tape recorder and the data received was downloaded on the ground and processed by the surveyors.

All the Tellurometers used Klystrons and it was starting to become old technology and tuning was a question of fine polishing of the klystrons and peaking the output.. To test and tune them properly it was advantageous to have a spectrum analyser capable of covering the Klystron's frequency range. I was asked to see if I could arrange to send one to Medan, I had the task of searching the various inventories and trying to find one that could be made available. Army did not have one, Engineering Development Establishment (EDE), a government organisation, was approached and they had one but were reluctant to loan it for an overseas deployment. RAAF had one and I was able to arrange an inter-service loan. EDE came back and said they were developing a solid state crystal oscillator that they believed was suitable to replace the Klystrons if we wanted to test it. Given that I knew nothing about it and the techs on the ground would know even less I was not keen. Finally it was agreed that I would fly to Medan, in mid-May, taking the RAAF Spectrum Analyser with me, and spend a week there trying to set up the testing regime we had discussed back in Melbourne. If successful we could, hopefully, improve the reliability of the data. Because it was a field deployment I travelled with an issue of field gear as I could find myself operating out in the scrub. My first thought was," Definitely no flash hotels or Magnavox expenses this time around, stretchers and sleeping bags under mosquito nets."

However, as I was given a single entry visa for Indonesia and could not fly direct to Medan I had to go via Singapore by Qantas and Singapore Airlines, to Polonia Airport. I was met at the Singapore airport by a Foreign Affairs official, who I had known as a WO 2 Chief Clerk at the School of Signals. He took me to a four star hotel and arranged for the spectrum analyser to clear customs and secured ready for the on-flight tomorrow.

After breakfast he picked me up and saw me through the Singapore Airlines departure gate to catch the short flight to Polonia airport.

I landed at Polonia to be met by the Operation's Admin Officer, an RAASC Captain Alistair Pope. Alistair and I knew each other from the TAC1 Course at JWC at the beginning of 1972. He was an interesting character, he was a linguist who spoke and wrote Bahasa Indonesian, and a chess master who wrote the chess columns for the Australian and the Canberra Times.

He cleared me through customs and immigration and made sure my gear and the spectrum analyser were taken to the camp. The camp was established in a kampong near the main airstrip. it had its own runway for use by the Operation's aircraft, the King Air that was fitted with Aerodist, an Australian Army Pilatus Porter a STOL(Short Take Off and Landing) aircraft, two RAAF Iroquois Helicopters and a RAAF Caribou that was used as a shuttle bus between Medan and Singapore. Apart from a couple of tin roofed hangars and few air-conditioned demountable workshop modules we were all housed in a series attap huts, similar to the one shown below, except ours were larger and raised about 200 mm off the ground. There were about ten all told with one used as a HQ office and OC quarters, one as a kitchen and another as an Officers' and senior NCO's mess. The rest were barracks blocks and I was bunked in with Allister, the Army pilot, a national serviceman named Moon but nick-named Face, our doctor the RMO and the unit paymaster/finance officer a CMF Captain from South Australia on full time duty. There were two other RA Svy officers but they had their own little hut.

An attap dwelling is traditional housing found in the kampongs of Brunei, Indonesia, Malaysia and Singapore. Named after the attap palm, which provides the wattle for the walls, and the leaves with which their roofs are thatched, these dwellings can range from huts to substantial houses. The diggers were quick to ensure they had as many of the comforts of home as they could and set up a very nice

little beer garden RSL by cutting a flap in the aback of one of the attap huts and arranging that a refrigerator and power to run it was set up. I think it was operational halfway through

48

day one. It was better to keep them close to camp rather than loose among the lovely ladies or 'shims' of Medan.

When the advance party had come to Medan on reconnaissance some time before the deployment they were assured by the mayor. "There is no VD in Medan" he proclaimed with hand on heart or, being Indonesian, rather an eye on the potential graft he could expect.

Unfortunately a couple of the advance party took him at his word and brought home an unwelcome memento of the trip.

Once the main body arrived, the RMO did some testing and his view was that anyone silly enough to avail themselves of the local females stood a good chance of a quick flight to Singapore for treatment for gonorrhoea. Most listened but some would never learn and made the trip more than once.

After I was there for a few days it was obvious that the TEGs were out of their depths without senior, experienced guidance and I flew back to Singapore to the HQ where I could use a phone to talk to my boss in Melbourne. Phone calls from Medan to Melbourne were next to impossible. He agreed that we needed to raise the bar with regard to support, of Aerodist in particular, so a couple of things happened: my week long stay was extended until the end of the operation, i.e. three months rather than one week, and John Strautins would come and join me.

EDE was again keen to trial their solid state oscillator and would send the design engineer with it if we would help him on site.

It was fine by me as it meant another electronic specialist on the ground and one that was used to using the spectrum analyser we had borrowed. So it was decided that

John would be there ASAP and the EDE engineer, complete with oscillator, would follow a week later.

There were a couple of problems with these changes. First Carol had to be told that I would not be home for a few months, someone had to arrange for me to be taken on strength so I could be paid, and how was I going to do my job including flying back and forwards between Medan and Singapore when I only had a single entry visa and had' used it up' when I first arrived in Medan. The answer was to ignore the problem and not include me on the manifest pretty confident that the rather lax immigration at Medan would not check who was or was not on our manifest. If sprung the obvious alternative was bribery a significant part of the local economy. Luckily we never had to resort to Plan B.

The OC was a fine man Hugh Taylor and he ran the operation with a soft hand but with a determination to get the job done. When I returned with the news of the changes he was a very relieved man as it appeared as if he might pull up short on attaining his targets if we could not improve the Aerodist results. He made a point of telling everyone on parade that I was to become a full member of the operational staff instead of just some blow in and we worked well together. Later on this was to be seen as a good move.

A few nights after I arrived, the RMO, the Finance man, Alistair and I went into town and had a break from camp.

We started with a walk to the main gate about a kilometre or so and there we were met by a group of young locals with finely painted and decorated 'betchas' an Indonesian pedal powered trishaw. The young pedaller sat on his bike seat whilst the passenger, or two, rode in comfort in the padded comfort of the carriage part.

The going rate to hire the betcha and driver for the night was 200 rupiah, and given that it was 660 rupiah to the Australian dollar it was a cheap way of getting around town. It has to be realised that a bank clerk did not earn much more than 200-300 rupiah a day it was a good return for effort expended. We each hired a betcha and set off for town.

The first stop was at the Hyatt Hotel where we had 'mee soup', a noodle soup with abalone. This was followed by a game of ten pin bowls, a first for me.

There was a casino attached to the hotel and you had to show about 5000 rupiah to get in, anyone with less than that could go into the cheap casino in the car park. Being white we did not have to flash the cash. We watched as a Chinese lady probably about 50 years of age sat at one end of the roulette table with two young girls acting as her bet placers. She would place bets of 10000 rupiah at a time and just point and directs her acolytes to place the bets with a wave of her fan. Money was certainly no problem. This was not a unique event every time we went there she was there and doing the same thing.

After a short stay we went to a beer garden and had a couple of Bintang beer. We had to withstand the assault by bar girls who would flutter around as soon as we arrived with," You buy me drink?" Given the RMO's warning we were easily able to resist temptation and escape. There was one funny incident a few weeks later when we were more use to the social scene.

We had become regular patrons going into town about every second night and the girls wanted to know our names so we established aliases, the finance man was Captain Money , I was Captain Aerodist, Allister was Captain Chess. One night I was sitting talking to Captain Money when one new girl turned up. We knew most of the regulars by now, but this one was very persistent and really wanted to lead me astray. She was becoming a nuisance but was impervious to hints. Finally one of the other girls decided to get rid of her, and really showed her displeasure. We were a bit amused by this and tried to quiet things down. The upset girl must have thought I was weakening and grabbed my hand. "Tedak! Tedak!" No! No! "Sama! Sama!" Same! Same! She said pointing at my crutch. The new girl was a boy, a shim in the local vernacular. I never suspected and could see how someone might be taken in. Allister had been, laughing to himself, listening to the chatter of course and knew what the girls were saying but let me find out myself. I like to think he would have jumped in if I had not got the message.

Allister was the perfect choice for this operation because your average Indonesian man on the street seemed to have only two loves, thick sweet black coffee and chess. Many were the times he would be seen sitting with some locals outside their front door, drinking the coffee they always offered guests to their 'front yard'. The front yard most often was a patch of dirt in front of their doorway where they would sit at night and smoke, talk, drink coffee and play chess. He would chat to them and challenge them to a game giving them his queen before they began. I suppose he might have lost a game but I never saw him lose.

Teams of two surveyors were deployed with a Tellurometer and a tent and stretchers by helicopter and stayed on site while the Aerodist aircraft flew laps between them. I forget how long they used to be deployed but while those Tellurometers were deployed we tweaked and tuned the spare ones back in camp.

They had the odd interesting moment during these deployments. One team were up in the highlands of Malaysia trying to anchor one end of the geodetic connection and reported during their daily sked that it was snowing. We were happy to accept, it might be cool, but wondered what they were smoking. A more credible story came from a Sumatran team who photographed some tiger tracks near their camp site. They were extracted that night.

With John and I both working in the repair shelter and the EDE engineer helping the TEGs become familiar with the spectrum analyser we were on top of the equipment problems and things were going well. I hitched a ride out to a visit to a survey team in a helicopter and had my first look at the surrounding countryside and the jungle and river systems.

The next tourist run I did was in the Army Pilatus Porter but we did not fly over Lake Toba (Danau Toba) that time, however, something was on the programme that was different.

When we came into land Face decided to demonstrate, without warning me, a beta approach landing. A beta approach is an interesting manoeuvre where by the pilot points the nose of the aircraft at the ground and dives before reversing the pitch of the propeller causing a braking effect. Just off the ground he levels off and lands, a very STOL landing indeed. It was an experience but not a pleasant one.

Danau Toba is the largest volcanic lake in the world and is a beautiful blue colour. The RMO was a keen photographer and grabbed every opportunity to jump on the Porter so when it was due to take supplies into a small airstrip up country he jumped on and convinced Face to fly back via Danau Toba. The rest of us were working quietly away then suddenly things got all tense, the Porter was overdue. It should have landed by now and we could not contact it. Hugh Taylor alerted Singapore HQ and a SAR alert was put out. All we could do was put the helicopters up and do a grid search on the route they should have been flying and divert the KING AIR to look as well. Time went by very slowly with no sightings, but eventually we received a phone call. They were on the ground in a field some kilometres away, the plane was damaged, but all on board were safe.

It turned out that the RMO had convinced Face to fly low over Danau Toba so he could get some photos and they flew into a power line that crossed the lake. The cable hit the tailplane damaging it and taking away the aerials but Face was able to limp to an open field and land. It had taken time to walk out and find a phone. Everyone was relieved but both of the miscreants were in deep doodoo.

Poor old Hughie had a mini collapse and was put to bed for a few days. The pressure of the early operational problems followed by this latest drama was too much. As I was the senior Army Officer I became the de facto OC while he was recovering. Not that that meant much as it was simply a 'what if' role in case of a drama. I had to handle some signals arranging for a crash investigation team to inspect the Porter but the RAAF pilots handled most of the details I just signed the signals. The crash investigation team include Maj Bob Millar and they went to work immediately. Eventually the Porter was brought back to Polonia airport on a truck and a Hercules flew it back to Australia.

The last few weeks of the operation were spent trying to complete the geodetic connection so John and I went with a survey team in a helicopter to a little island paradise on the Johore straits. We arrived early and the surveyors set up their gear while John and I had a swim and explored. Just over from where we landed was a fishing kampong, two long houses with jetties extending out into the straits. The fishermen were pleased to have company, even if we could barely communicate, and showed us their boats and nets and gave us an impression of how they fished. To show how rich with sea life it was they cast lines off their jetties.

Their rods were thin bamboo or reed and they could whip a lure a fair distance with them. On the recover you could see large garfish chasing the lure. They never missed catching one but we did not have the knack because the garfish would look at our casts and shrug and swim away.

We swapped a packet of Marlboro for a few of the gar and cooked a BBQ on the beach while the Surveyors surveyed. It was a good relaxing day and some good readings were recorded so everyone was happy. One of the ironies of the trip was that the RAAF having dropped us off flew about 500 metres away and landed on a different island and had a private RAAF only picnic. 'Strange cattle these RAAF'.

Eventually like all good things we had to pack up and return to civilisation. The bulk of the Svy Squadron were flying back by RAAF Hercules but as John and I still had the return part of our commercial ticket left we flew back to Singapore via Singapore Airlines and transited straight onto Qantas First Class for the flight to Melbourne.

Sometime later I read where all soldiers who had served in Indonesia during a certain period of time before 1975, and on Operation longer than three months in duration, were awarded The Australian Service Medal 1945-1975. I asked if John and I had been recommended for the medal but of course we had been overlooked as we were there on detachment not posting.

It took a little bit of effort but I was able to show through our pay records that we were on the 5 Field Survey Squadron strength for the operation and we were both later awarded the medal.

Once we were back in Melbourne John returned to his posting in the Technical Support Unit at Broadmeadows, the unit required to conduct investigations into equipment problems determine how to repair and maintain equipment, and to write and publish the Electrical and Mechanical Engineering Instructions (EMEIs) to explain it all. I similarly went back to DEME at Albert Park and settled down to doing the task I was supposed to do, rather than cavorting all over the place.

Time was once again our friend as we gained enough points to qualify for a MQ this time in Watsonia amid a group of fellow officers who worked at Albert Park Barracks. This meant we could car pool and it was a very relaxed way to get to work after such a long time driving from Frankston to Albert Park.

On 1 November 1973 there was a change in the army organisation and DEME, or our part of it anyway became part of a new formation called Logistics Command and my posting was amended to be Staff Officer Grade 3 Electronics. It did not change anything at the time but would be relevant later.

Chapter Five - 101 Field Workshop 1974-5

Army postings tend to be for two years and my time in Albert Park was to come to an end in March 1974, with a posting to 101 Field Workshops at Ingleburn, NSW.

To complete the move I left earlier than Carol and the kids to arrange a MQ and prepare everything including school for Keiran. I decided to drive the car and loaded it with my uniforms and some civil an clothes and of course my sports gear and drove to Albury where I overnighted and continued on the next day Wednesday 27 March 1974. I drove in to the camp area about 1400 hrs but no one was in HQ. It was a sports afternoon and everyone was over at Holdsworthy where '101' were to play a rugby scratch match against a team from 2 Base Wksp Bn.

I couldn't wait and drove over straight away and arrived just on half time. Impatient to meet my new colleagues I changed into my gear and ran on, no warm up. I lasted one decent sprint before I tore both hamstrings.

The moral of the story you cannot drive five hours and run on to a rugby pitch without warming up. I had a haematoma down the back of both legs from bum to knee for over a week and had daily treatment in the Ingleburn Area hospital. Not quite the entry into unit life I had hoped for. When the game was over I sat with ice on the back of both legs and talked football over a few beers. It was here that I met the other officers of the unit. Some I knew already, Maj Bob Millar who I had last seen in Medan was the OC, and Capt Peter Lawrence my golfing and rugby mate from WA was OC Vehicle Platoon. The three others I had not met before, Capt Bryan Coolahan was the Adjutant, Lt John McNamara. a Royal Australian Ordnance Corps officer was the OC Stores platoon, and a newly graduated 2Lt Peter Malone was the OC Recovery platoon. I was to be the OC

General Engineering (GE) Platoon and, as the senior captain, the 2IC. The GE Platoon employed all trades other than Vehicle trades, radio, electrical, fitters, and welders. I have mentioned earlier the role of a Raeme field Wksp and to perform this role, we had five officers and about 100 other ranks.

Over the next couple of weeks I limped around and recovered slowly but was able to find a MQ not far from the unit, and close to the hospital and the local primary school, and arrange the removal of family and goods and chattels. Carol and the children were still in Watsonia but, when the removal was on, I flew back to supervise the loading of our furniture etc. After it had taken off on the truck I took Carol and the kids to Tullamarine where we all flew to Sydney and were taken to a motel on the Hume Highway just near the turnoff to Ingleburn. This was to be our temporary home until the removal truck arrived and we could occupy our MQ. The MQ was a 3 bedroom timber place that was one of a number of pre-fab houses bought originally, I think, for the Snowy Mountain Scheme but now used by the services as MQ. Some of these were made in Sweden and some in Austria, I think ours was an Austrian one not that it mattered they were only nine squares and were adequate at best. We had the main bedroom, Keiran had her own room and the boys had wooden bunk beds in the other bedroom.

Ingleburn was a small village about 3 km from the camp and any shopping had to be done in Liverpool or Campbelltown. We chose to go to Campbelltown as it was a quieter area than Liverpool.

****

My first trip into the bush was a small deployment of a Forward Repair Group (FRG) and it was on the Holdsworthy range. It was a teaching exercise, for the new march-ins, me in particular, and was only for a week. Seeing that it poured

[Type text]

rain all week and the soil turned into yellow, slimy, oozing clay, the main benefit was that gained by the recovery elements getting the workshop vehicles out of the bog and on to hard ground.

Still I developed a rapport with the Senior NCO on the exercise a staff sergeant who was actually in the Vehicle platoon and worked for Peter Lawrence. Whatever it was we clicked and for the next nearly two years we remained the best of friends and he came with me on many similar exercises.

101 Fd Wksp's role was to provide field level repair support to 1 Task Force (1TF) units that did not have integrated workshops and/or to take any overflow from those workshops. (See previous explanation of levels of repair, pg. 259.)

When deployed in the field the workshop could exist as a single unit but this was not a tactically sound method of deployment, given the difficulty of self-defence when the majority of the unit are working at their trades and not available to man any defences.

By 1974 in the post SVN days, when military thinking was swinging towards conventional warfare and defence of mainland Australia, we had to develop and/or relearn skills that had not been trained for, or practiced, since WW2 or Korea. These skills involved the tactical movement of the unit over long distances, and when in the operational Area, to deploy the workshop as part of a Task Force Maintenance Area (TFMA) that included an Ordnance Supply Company and a Transport Squadron and other minor logistic units. The Exercise programme for the next two years included two task force exercises and a divisional sized exercise. The last one of those had been in the early 60s about 1962, i.e. pre SVN; and had not included the deployment of the logistic support units in a TFMA.

[Type text]

Our initial training exercises were independent of other units, simulating a TFMA deployment, but with the other elements merely circles on a map overlay. To give us a workload we worked in concert with 2 Cavalry Regt who like us were working up and developing their own response to the new concept of threat to mainland Australia. We established a good working relationship and worked together twice in the hills behind Nowra.

I found during these exercises that it was possible to do the job, without being silly about it, and as we always had generators to provide power to the workshop vehicles I always carried the old 21 inch TV in my vehicle and trailer. To better explain what I mean I should mention that all the trades vehicles were Landrover with a half-ton trailer that could be joined together with a canvas cover over the lot making a roughly 12 ft by 10 ft working area that could be folded up and packed away and moved to a new location in a very short time. The Officers' vehicles were a Landrover with quarter-ton trailer that carried a 11ft wide x 22 ft long tent, stretcher, chair and table/desk. Thus they were both office and home from home. Each evening after the evening meal the CO and Adjutant of 2 Cavalry Regt would drop in for a coffee and watch the news. This gave us a nice relaxed way to prepare for the next day's activities and to sort out any problems.

Sport was, of course, a special part of unit life and we had a good team with a few ring ins from the smaller units, e.g. the MPs and the army reconnaissance flight. They were too small to field a team of their own but were a welcome part of our team. Our competition were the 'reserve grade' sides of the larger TF units, 2 Cavalry Regiment, 5/7 RAR, 8/14 Field Artillery Regiment and 1 Field Engineer Regiment, and 104 Signals Squadron. a unit our size. We could hold our own with most of the opposition, except for the Artillery team that was always too powerful for us. Four of the workshop Officers Peter, Bryan, John and myself played as did two of the pilots (both captains) from the recon flight so it was the perfect set up for an army side, a good mix of officers and ORs.

Two memories of this competition are worth recalling. Firstly, the captain and front row prop for B company 5/7 RAR was to go onto make a name for himself, Sir Peter Cosgrove, I can honestly said I have belted and been belted by a knight of the realm. The second was less notable but made an impression on me. I went down on a loose ball and collected a kick in the ribs breaking two and collapsing my lung, a pneumothorax, and another visit to the Camp Hospital.

****

After a couple of the independent exercises the unit became very skilled at packing up, moving, deploying, absorbing workload and then doing it all again. By October we had it down pat and just in time for a major task force sized exercise near Broken Hill, it was called Exercise Boundary Rider, and would be some 5-6 weeks long, with a week to get there and back.

To add to the excitement, Maj Bob Millar, the OC, was posted just before hand and I became OC with effect, 10 October, and was responsible for taking the workshop on its first serious deployment for many years.

I forget how many vehicles we took with us but there were certainly forty to forty-five so it was a major part of our planning just setting up the convoy in discreet packets with our integral recovery vehicles on hand in case of accident or breakdown. We went from Ingleburn through the Great Dividing Range near Bathurst to overnight at the Cobar showgrounds and then into the Broken Hill area near Menindee Lakes, a chain of freshwater lakes, near the town of Menindee, that supply water to Broken Hill.

The trip down the Barrier Highway was fraught with danger as very few vehicles out there travelled at less than flat-out so if you had an accident with a civilian vehicle it would be a bad one. The other problem was stopping at the side of the road for a rest, brew etc.

We were told at the briefing that we could not stop at Menindee as the indigenous inhabitants would seek to steal anything not tied down.

The other problem was that a fully laden Workshop Landrover, was very heavy and took a while to stop given their drum brakes and we were likely to meet up with kangaroos, wombats etc. at night so we could only drive by day and even then carefully. To illustrate the point we came across a Torana that was obviously moving when it encountered a wombat. I do not know about the wombat but the driver was killed. It sobered up the more lead footed of our drivers.

Bryan Coolahan was the one responsible for the route reconnaissance and convoy orders and went on ahead from the lake to secure and mark the area we would occupy in our initial deployment. It had been my job as 2IC to go with him and do the actual siting on the ground where the workshop vehicles would go then call forward the OC to confirm it and complete the deployment. In this case as OC I just went forward and did it in one step. A workshop fully deployed with all its vehicles, kitchens, stores and repair parks (where the vehicles go before and after repair) occupies a grid square or a square with sides 1 km long. A ring-road runs off the Main Supply Route (MSR) and loops back to meet the MSR again some 500 metres down the road.

The weather at Broken Hill was perfect and we had a solid workload that we were able to process and return to the units in a reasonable time. Some TF units, however, stockpile their equipment needing repair and bring it with them, knowing we will do our best to turn it around rapidly without it being on the end of a queue as could happen back at Ingleburn where other priorities may interfere. The logic is, out in the bush they are our primary concern, and the Comd 1TF is watching to make sure his units are all ready to perform.

Of course that means a large initial workload that stopped us giving service to exercise-caused repairs, slowing them down until we cleared the backlog of old repair items. I spent many hours of my time trying to make the quartermasters of the units appreciate that if they gave us the repairable items before we left and marked them, realistically, as needed-on-exercise we would do everything to have them back to the unit before they deployed. I must not have an honest face because few seemed to believe me and on the next exercise the same thing would happen.

After our first deployment we stayed in place for a week before having to do a night move to a new location. I was in the first vehicle and missed the small sign that would put us on the right track which was to be the new MSR and led everyone in a big circle- lost! I finally gave up and set the Wksp into a temporary location waiting for daylight. Very embarrassing. Early, first light, I climbed on top of a truck to see if I could identify anything on the ground that I could place on the map and get back in the war. I think we may have been behind the enemy at one time!

I had to report my location to 1TF HQ and this was not a radio call that I was looking forward to. Initially I was talking to one of the staff officers and he asked if I could see anything and I said there is a dam and a windmill about I km or to the south of me. "How big is it? "This clown asked.

[Type text]

Considering that nothing out there was of any size, I got facetious and said sarcastically, "Nothing to write home and tell your mother about."

Next thing the Comd 1 TF, Brig Leary himself, was on the radio giving me a boot and telling me he could not care less what I had to tell my mother, just get moving and get to the correct location ASAP. Luckily the patrol I had sent out found the markers and we were able to move immediately. I kept out of Brig Leary's way for the rest of the exercise.

****

We had a break from exercises until after the September school holidays so we loaded the kids in the back of our 'new' ford Escort and drove to the army holiday accommodation alongside the Coolangatta airport. They were three-bedroom fully self-contained units with a full sized pool and only a short stroll to the beach, Coolangatta, and Tugun we loved it and it way our regular annual break away for a number of years.

Although we spent nearly 80% of the year on exercise there were still some relaxation and fun the majority of this was mess-based but occasionally we would lurch out onto the wild streets of Kings Cross. These excursions were generally celebratory. Peter, Brian and I were doing our major's exams and John was doing his captains, so when the results came out and we had all passed those we had attempted, we headed off to the Cross.

We went in and were picked up by the duty driver in our staff car so no driving dramas; it was a splendid and safe night. The next year when we had all finished successfully we repeated our new tradition.

Bryan had obviously not had enough of the bush or the Colo river area. For he arranged an Adventure Training exercises where by half the unit would go by vehicles to a start point about 12 Km down streams from Wiseman's Ferry with a number of canoes in the trucks. The other half of the unit would walk from Wiseman's ferry to the Start point and would ride back in the trucks. Our half would paddle to Wiseman's ferry and we would have a day or so relaxing and canoeing on the Hawkesbury, as a reward for a hard job well done.

There is an old army saying, 'Time spent on reconnaissance is never wasted!'

I unfortunately did not remember that in preparing for this little paddle and thought (assumed) the river ran from Start point to Wiseman's Ferry i.e. we would be going with the current. It did not take long for me to work out that I was mistaken and it ran the other way. Itwas a hard slog and I have never been happier than when we finished and went to the Wiseman's Ferry Hotel.

Someone had a smart idea for a drinking competition, I forget who, but will blame WO1 Bob Billet, our Artificer Sergeant Major (ASM) and the senior soldier in the unit. He was good man and it was a privilege to work with him although we did occasionally have a different point of view on some subjects. This, however, was not one of them as I stupidly agreed to participate. The rules were simple the barman produced four liqueur glasses and filled them with Cointreau, the top was lit and while it was burning you would bend over, put your mouth over the glass and skol it. You could not pick the glass up by hand that was for pikers. Look Mum No Hands (Or brains.) We played this game far too long and by the time we went to bed we were very nicely thank you. Later the next day when we went back to the hotel for a quiet drink the same drinkers from last night were there plus some of their mates hoping for an encore. I recall doing a demo heat but that was it. I have not been involved in this 'game' since.

In November after the exercises were finally done and we were back in camp for a while it was decided we should do a full mess night in the Cross so we all, including wives and girlfriends, jumped on an army bus and went to a theatre restaurant,, the theme was the 90's and we fitted in in full Mess Dress. I was still OC and Bryan or John or both had set me up to be called on stage and stand like a 'dill' while a heavily made up women who could have been my mother sang to me, something like "All I wanted was a soldier." This soldier did not know whether to join in and risk having her belt me for ruining her act or what to do so stood there and copped it. When we were walking back to our bus that was parked in a floodlit car park we walked past a parked Minivan that was bouncing on its suspension.

Fascinated we went to check this phenomenon out and looking through the window saw a rather violent business transaction being played out. We gave them both points for style and left them to it

Chapter Six – Cyclone Tracy

Christmas 1974 was to be a major event at Baileyana Street with the now three grandchildren to be spoiled. I had become so used to sleeping in an '11X11' tent that I threw one on the roof of the escort and took it to Frankston for extra accommodation. We drove all day having left about 0400hrs Christmas Eve and reached Frankston, 900 km away about 1400hrs. I found that I could usual average 90 km per hour over the trip including stops. While everyone was inside playing grandparents I set up the tent and was ready for a good Christmas break. Then someone opened a door up north and in came Tracy.

Cyclone Tracy was a tropical cyclone that devastated the city of Darwin, from Christmas Eve to Christmas day, 1974. It is the most compact cyclone or equivalent-strength hurricane on record in the Australian basin, Category 4 winds on the Australian cyclone intensity scale, she killed 71 people, caused A$837 million in damage (1974 dollars) destroying more than 70 per cent of Darwin's buildings, including 80 per cent of the houses, thus leaving more than 41,000 out of the 47,000 inhabitants of the city homeless. This necessitated the evacuation of over 30,000 people to the south.

You did not have to be a rocket scientist to realise the army was going to have a major role in the rescue/recovery that would follow so that finished our Christmas and we packed up the tent and loaded the kids back in the car for a 0400 hrs start on Christmas day to return to Ingleburn. Once again we made the trip in 10 hours and I was in my office that afternoon. I had rung the Duty officer from Frankston and told him I was on my way back and to recall as many as we could from leave.

The next day I attended the conference at 1 TF and assured the Comd that we would be able to move out rapidly if required. We also had some barrack lines that could be made available for the refugees if needed. Bryan was organising the cleaning of those lines as we were in conference. I expected they would ask for a Forward Repair Group with me taking it but that did not happen.

When the reconnaissance was done by Maj Gen Stretton and his staff they decided they only needed our electricians and all our generators and no officers required as most of the force was coming from Townsville. So there I was all dressed up but nowhere to go. Similarly the lines that Bryan had made ready were not required either, so we switched off and sent people back on leave.

Chapter Seven - TFMA Exercise

In early January 1975, Maj Fred Barlow arrived to take over as OC and I slipped back to OC GE Platoon and 2 IC again.

This year was to be a big one for exercises with another task force size exercise at Broken Hill and a divisional sized exercise near Singleton in central NSW. After the previous year we worked like a well-oiled machine and Fred was given a free ride as OC. He had served with 106 Fd Wksp in SVN but this, defence of mainland Australia concept, was new to him. We worked well together and the transition of command was smooth. He did not try to do a 'new broom' job on us but slowly impressed his own personality and style on the unit.

The Broken Hill exercise went well at least I did not lose the Wksp this time which was a relief and we had a few more Nowra exercises with our old friend 2 Cav Regt to tune us up for the big one at Singleton.

A divisional exercise is a huge thing and this was the first for a long while so we were all keen to put on a good show. For the logistics units it was to be just that 'a show'. There had not been a full deployment of a TFMA in living memory and everyone and his dog wanted to visit it and see it on the ground. Therefore, part of our responsibility was to expect hordes of visitors and we should prepare a brief which was to be given verbally to the visitors as part of the tour. This was as well as doing our normal tasks. When the TFMA was on that ground it was so large that visitors could only see it by bus. They certainly could not cover it on foot, so we had few visitors other than other Raeme people, but did see a lot of bus traffic.

****

Prior to the big exercise we had all sat the remaining promotion tests and had all passed and were qualified for our next promotions. Naturally as I mentioned earlier we celebrated and relaxed until the powers that be, the military secretary (MS), found the time to issue a posting order.

Our move to Singleton took us NW out of Sydney into the Hawkesbury area, not far from where we had our Adventure Training, and the approach road the Colo-Putty Rd was narrow and winding. We made it alright but had only just settled in and we had to send our Recovery Platoon back to recover a truck that was carrying an M113 Armoured Personnel Carrier that had gone off the road and down an embankment. The recovery guys love this stuff and they were able to demonstrate all their skill and training in this instance.

The M113 was still attached to the trailer and apart from a list to port the damage to the load, prime mover and trailer was limited. The SSgt Recovery laid out his tackle and set up a brace that ensured the load did not fall off or the trailer and load slide further down the embankment. With the job secured they set to work swinging the front of the truck so it was pointing downhill they worked the trailer around so the M113 could drive off. Once the load was removed from the trailer they set up a tow and pulled the rig onto level ground and the operator's drove the M113 and the truck and trailer cross country to a convenient track.

The silly part of this was that the young transport officer who had overseen the tying down of the M113 on to the trailer was charged because he did not use the accepted tie down method. Seeing the load stayed on the trailer we thought he was harshly treated.

After the exercise had finished we moved out to a non-tactical area on the Singleton range and had a BBQ. No booze just a BBQ and some home grown entertainment. It was all done with taste and decorum and for the final act we, (Bryan, John and I), gave a rendition of 'Old MacDonald's farm with appropriate gestures. When it came to the final stanza of the 'bears are bearing it now' we spun around and backs to the audience dropped our trousers. The diggers thought it was the best thing they had ever seen. Three of their officers prepared to join in the fun.

This turned out to be only the first performance in a series. The following Saturday night all the Officers and SNCOs were invited to the Singleton RSL. Late at night, just on closing, we were pressed to do an encore. It was such a success that we were invited back the next morning to do it again this time with a full RSL as audience, we were getting pretty good by this time, and we brought the house down. Needless to say we could not buy a beer all day. Seeing it is now over 40 years since then I wonder if anyone in the RSL remembers

****

After we settled back into Ingleburn it was time to clean all the entire vehicle fleet and repair some of our own equipment faults and take advantage of the break to send the diggers on courses particularly promotion courses that they had not been able to go to given the bush time we put in. Similarly, I went to RTC to complete my Corps qualification for promotion to field rank, i.e. Major, from 3 September -30 September. Unfortunately during the period of my time at RTC I made the biggest mistake of my life when at a party I fell for a divorced woman Jan. She was a UK born Jewess who fascinated me and I forget all that I had with Carol and our family and made a fool of myself by leaving them for a time in pursuit of someone who did not see me in the same way that I imagined her.

Chapter Eight – Leopard TFCS

It was during this quiet time that my posting order came through. I was posted to the Technical Support Unit (TSU) in Broadmeadows, Victoria as the OC Electrical and promoted to temporary Major. I would take up my posting on 17 November 1975 and be temporary in rank until 30 June 1978. TSU was the premier technical unit in Raeme and was responsible for the evaluation of equipment repair and maintenance of all Army electrical and mechanical equipment and the production and publication of EMEIs for the units and their repair facilities. As OC Electrical I supervised the work of about eight very experienced warrant officers or senior Sergeants all of whom had qualified as artificers in their respective trades. As I mentioned before the, qualification, artificer, shows you to be the best of the best in that trade.

When I took up my posting I had moved Carol and the kids to a MQ in Watsonia and just up the road were Ruth and Barney Goodchild our neighbours from WA. Barney and I played golf and we began a regular series of family nights playing canasta and having fun. Early in 1976 still working at TSU, I was about to make a big mistake and also be offered a great opportunity. The first was, as I alluded to earlier when, I left home and moved into the Broadmeadows Officers' Mess, believing that my future was with Jan, but she was just playing with me and soon I found myself sans family and sans girlfriend.

The opportunity came as a result of the Australian Army beginning a trial to determine which tank would replace the obsolescent Brit Built Centurion. The contenders were the US made, M1 and the West German (FRG) made Leopard 1 and a team of Armoured Corps and Raeme officers and SNCOs had spent some years conducting trials before selecting the Leopard 1 as the winner.

The illustration below shows the Canadian CA 1 version but apart from some minor details is identical to the Australian AS 1 Leopard.

The Leopard 1 was a European build rather than just an FRG build. the hull complete with engine drive and tracks were made by the overall prime contractor, Krauss-Maffei in Munich, the turret by Wegman in Kassel near the East German border, the tank fire control system (TFCS) by SABCA in Belgium, the laser rangefinder by Selenia in Rome and the main armament was a UK built L7A7, 105 mm Rifled barrelled, crew loaded gun. The only US made part was the Cadillac-Gage weapon stabilisation system assembled in Mosel FRG.

B ridge layer

Engine
er Vehicle

Armou red Recovery vehicle

The most telling reason to choose the Leopard1 over the M1 was the family of vehicles that came mounted on the same hull with the same drive and track system, see example illustrations above. The logistics advantages of a fleet of family-vehicles instead of a mix, was obvious and swung the decision Leopards way. (Although Tank vs. Tank there was little between them.)

From a Raeme point of view the mechanical and armament were straight forward but the TFCS was a very different matter. The TFCS was based around an analogue computer with inputs: range to target, temperature, atmospheric pressure, wind direction and speed, ammo type and barrel wear. All of these inputs were processed by the computer to solve what is called the ballistic equation to improve the first round hit capability.

Let me explain the range to target is needed to compute the trajectory of the round, the environmental conditions affect the flight of the round, each ammunition type varies in propellant and shape so will fly at differing speeds and distances, and a barrel wears as a function of the number of rounds and the type of rounds fired, and that wear affects the muzzle velocity and stabilising spin of the round.

Australia's armour did not have anything of this sophistication previously so we had to do a Maintenance Evaluation to determine how we could repair and maintain the TFCS. It was a TSU task so WO1 Stretch Woodhouse and I prepared for the Grand Tour I had been offered a choice firstly do the ME, or go to Bandung in Java Indonesia and do a two year posting at their version of RTC. An Albury manufacturer had developed and made a target system called DART and the job would be to instruct on this system. Given that I was betwixt and between with my family and social life I did think about the posting but elected to do the ME.

In late1975, although it had not concerned us at the time, DEME moved from Albert Park to become part of Logistic Command, but TSU stayed out at Broadmeadows. To make things easier when organising the ME, Stretch and I moved into a new office in Log Comd on the18th floor of the Defence Centre a new building at 360 St Kilda RD, Melbourne. From here we had access to more clerical assistance and were better placed to process our visas and security clearances.

Because we were visiting some high security areas and affecting ministerial decisions we had to have a particular clearance. We filled in the forms to gain these clearances and when the Tank Project Liaison Officer, (TPLO) Bonn asked if we were cleared I asked the security guys, and when told yes sent the answer to the TPLO. I later had an angry security officer telling me I was not allowed to know my own security clearance?? I could not and still cannot, understand the logic but shrugged and continued on making our plans and bookings.

The tour started in Bonn then Brussels, followed by Leopoldsburg in Belgium then on to Munich, followed by Rome, then Kassel, and finally Munich again. We left Melbourne on 18 April 1976, and returned on 25 June 1976.

Because we were going to be spending some weeks in the tanks learning to operate the TFCS we were issued tank suits. I chose the standard version but Stretch scared of the thought of a cold European spring and start of winter chose the winter weight version. Unfortunately for him, 1976 was, quite simply, very hot. The mildest spring ever and he suffered from the heat of a tank on the range all day in the sun. After a short while he chose to wear greens and forego the tank suit. He wore his Peak cap at all times but I had thrown my bush hat in so wore that. The beret, my most preferred head dress, is only worn by mounted troops and that right was jealously guarded. Because we would be wearing a variety of uniforms and civilian clothes I went to a second hand store where I had seen a cabin trunk that had a lift out tray that allowed me to pack my service dress uniform flat and bought it. Therefore, on arrival it was just a matter of hanging it up and not having to worry about ironing. The resident Australian army reps in the TPLO in Bonn were RAAC, a Lt Col and a Lt, so I did not seek permission to wear the beret. I regretted it but my little bush hat fascinated the Belgians. They called me "Major en petit chapeau".

We left on Saturday 17 April 1976 and flew Qantas from Melbourne to Frankfurt via Sydney, Singapore, and Bahrain, and enjoyed the first class service once again. It was an eight hour flight from Sydney to Singapore and we were served dinner on the way with Sydney Rock Oysters, Cold Poached Salmon Alaska with a tossed salad, then a Pavlova and cheese and Coffee. Naturally we had our choice of French or Australian Champagne and finished with liqueurs. The stop in Singapore was long enough for us to go for a walk before boarding for the second leg, seven hours thirty minutes; they must have thought we were looking peaked as they gave us a champagne supper then breakfast before landing in Bahrain. From Bahrain we flew the five hours 40 minutes to Frankfurt enjoying a brunch along the way. I have no idea what the calorie intake was on this trip but it was

[Type text]

substantial. But I do know that when we arrived early morning Monday 19 April 1976 we were well fed and watered. We made the connection to Cologne from Frankfurt but only just due to us having to find our way around Frankfurt airport to the departure lounge. It would have been better if we had a guide to meet us or more time, but we made it.

An Austrian born RAAC Officer, Lt Wolff Klimisch, who was the Assistant TPLO, met us at the Cologne airport and took us by taxi at a cost of Dm40, to the Steigenberger Hotel (Room 728) on the Rhine. The room costing Dm 65 per day was not air-conditioned and as Bonn was very humid it was not as nice as we hoped. There was a black and white TV but as it was all deutsche sprechen who cared.

There was a bowling alley with restaurant as part of the hotel and after a rest in our rooms Lofty and I met there to have lunch the beer cost 80pf about A$0.25. To get with the local scene I had a wiener schnitzel, considering how well we had been fed on the planes it was too big for me and I could not finish it.

The manager of the bowling alley spoke excellent English and we enjoyed his company for the rest of the evening. He was a snuff user so naturally I had to try it. Okay been there done that, but was not tempted to make it a habit.

We had a look around and saw that there was a sauna and a swimming pool in the hotel. You had to read the times for the sauna as there was Damen only, then Damen and Heren, but no Heren only? Considering that there was a U-bahn just outside the door we found out how to use it as it appears the best way to get around. Being jet lagged we had a supper about midnight, spaghetti mit fleischsauce then collapsed on the bed at 0120 hrs local.

In the morning after a swim in the pool we walked in to the platz in Bonn and had a quiet lunch, the day was just like the

79

representative photo below. In fact I think we sat at the table shown. Opposite the statue of Ludwig van Beethoven and outside Stern Hotel. courtesy of Wikipedia)

To add excitement and amusement to a pleasant day Bonn put on a couple of vaudeville acts. There was a one man band, then three Germans dressed in Scottish national dress, including kilts, singing Hillbilly songs and as if that was not enough, the next act was a police car sweeping into the platz and stopping just down from where we were sitting.

Three very young policemen who looked like children in fancy dress then tried to arrest a large fat very happy drunk. He saw them coming and laughed and yelled "Schwein conmmen" much to our amusement. Finally he smilingly went with them.

It may be surprising to hear but we did have a few beers with our *wurst* and that convinced me that I was going to thoroughly enjoy my time in FRG. The rest of the afternoon we strolled around the pedestrian ways of the market and I bought a suitcase, socks, and a camera.

Shopping in Bonn

Wednesday 21 April 1976 was another good looking day apart from the standard haze that seemed ever present.

We were up at 0600hrs local and had a swim and organised our uniforms and some laundry. We confirmed that we would be checking out The next day had our breakfast early and were ready when Wolff came to pick us up at 0845hrs and drove us to the TPLO offices at Bad Godesburg.

The TPLO Lt Col Bernie O'Sullivan had made all the arrangements with those we had to visit and a rough itinerary was ready for us. Three names were to be of importance to us, Andre Scheitler, the Belgian engineer who worked for Sabca, the Belgian Aeronautical company that were the prime contractor for the TFCS, and who had assembled the TFCS into a working system and who would teach us how to operate and repair it, Dirk Hansen the Krauss-Maffei representative who had worked with the Tank Project team during the evaluation phase and Tony Hughes-D'aeth, the Australian Government representative, working with the prime contractor, Krauss-Maffei and living in Munich.

81

The brief on local conditions reminded us to always carry our passports on us, do not talk about what we were doing, keep away from East German border, and do not trust the telex system; send any reports etc. via the embassy when passing through.

Lunch was at a small place near the embassy and we all had steak and mushrooms and 2 stubbies of beer each for a total of Dm 94.50. The rate of exchange was 3 Dm to the A$.

Back at the embassy we picked up our tickets for the rest of the trip including rail from Brussels to Kassel. With Wolf as our guide we tried out the u-bahn back to our hotel. It was easy you stepped on board and put money in a slot in the carriage and it gave you a ticket what could go wrong with that simple system.

That night we were picked up at 1800 hrs to go to Bernie O'Sullivan's house for dinner. They had a nice house with a lovely view down the Rhine valley. Unfortunately I was coming down with a cold so disprin and bed was the order of things

The next morning I was still suffering from the cold with a sore throat, but after breakfast at 0700 hrs I went and had a haircut. A bit more than I was used to with a shampoo, head massage, haircut and blow wave. It cost Dm 18 or about A$ 6 which was again a bit more than I was used to.

Today was the day when I had to get my diary sorted out re timings of itinerary etc. and prepare to book out at 1130 hrs then lunch and a further briefing with TPLO. After lunch we flew down to Munich where it was cold and overcast. To familiarise ourselves we went for a stroll after booking into Eden Wolff hotel which is opposite Hauptbahnhoff and walking distance to Marien Platz. The room was small but comfortable, no TV but US forces FM radio. It cost Dm 76 per night. Later we met the people, and their wives, who would be most important in us completing our mission. Ziggy Ziekwolff, the Australian liaison Officer, Dirk Hansen a happy Bavarian, the Krauss-Maffei warranty liaison officer who was going out to Australia in August in time for the arrival of first tanks, and Tony Hughes D'aeth , it turns out that he also went to CBC Fremantle and did electronics afterwards, coincidences! Our reporting system was amended due to the problem of sending classified mail around Europe.

It was now Friday 23 April 1976 and our first day of actually doing our task. It was 4 degrees Celsius and snowing and after breakfast we took a taxi to Krauss-Maffei for a briefing.

Because of some Krauss Maffei problems the itinerary was reorganised and our programme was discussed in detail, item by itemed. It was interesting but all concerned were determined to make the ME work.

Finally the following was decided, our ME was extended in length by a week, later it was further extended by another four days making our RTA 24 June 1976,and I was appointed as warranty liaison officer (WLO) as it was considered that most of the problems were more likely to be electrical than mechanical

Having concluded our meeting we walked across to the Direktion casino the company senior engineers' mess hall. The lunch was soup, fish and salad and desert with a nice white wine.

Now it was time to visit the factory production line and we saw AUS 17001 the first Australian tank hull moving along the line with all suspension and ancillary bits being welded or bolted on. At the end of another line was a Danish tank complete with turret having its final fit out. Our next stop was where the Aust training cabins and driver simulators were being worked on. The last stop was afternoon coffee back in the briefing room where we made arrangements for the next period of time in Belgium, before taking a taxi back to Eden Wolff hotel.

Tony and Sue Hughes D'aeth picked us up and we went to their home for dinner, a very good night. On the way down in the lift I met a chap who spoke English, Bob Ross, who came from North Melbourne. I sat with him in the bar waiting for Tony and we agreed to have a drink together tomorrow night. Unfortunately my cold was back and I had to go back to the hotel early from Tony's.

It was overcast and drizzle 2 degrees C, the next day so I shouted myself a breakfast in bed. There was no laundry service on the weekend so I did my own and turned the bathroom into a drying room. Lunch was taken on the stroll with a bratwurst from a stall in the shopping area beneath the Marienplatz before coming up to the surface for a beer at the Ratskeller where Hitler began his career, just near the Rathaus-Glockenspiel a tourist attraction in the heart of Munich which dates from 1908. It was 1100 hrs and I was just in time for the show. It consists of 43 bells and 32 life-sized figures, showing a joust with knights on horseback representing Bavaria (in white and blue) and Lothringen (in red and white). Bavaria wins!

Returning to the Hotel I stopped at the Hauptbahnhoff cafeteria and saw what I thought was half chicken, 'halfe deutzen schnecken" it turned out to be a half dozen snails. Back in the hotel bar with Stretch we ran into the Australian from North Melbourne and had a drink. Later we were joined by two couples from Darwin. The Eden Wolff became the Aussie bar and as the next day was Anzac Day we tried to sample all the schnapps on show. Big mistake as I found in the morning.

We flew to Brussels via Frankfurt but missed our Belgian escort, Major Brunin due to a mess up with our luggage. We took a taxi to the President Nord on Boulevard

Napoleon Bonaparte, a quite nice hotel and only A $ 20 per day.

Deciding to play tourist we roamed around to find street upon street of restaurants. Each street seemed to cater, (pun?) to a different national cuisine. We chose well with the menu du jour for A $ 5, a beautiful soup, incredible coquilles St Jacques, an ordinary filet of veal and a pleasant fruit salad. Coffee and beer were extra but it bode well for the rest of our stay.

Maj Brunin had left his private phone number at the desk so we made contact. Maj Brunin was the epitome of a European army officer, smooth and knowledgeable and was able to speak fluent Flemish, French, Dutch, German, Italian and English and probably others I did not know about. At this time we backward ockers could not even order a beer without thinking about it and miming as well.

The next day Maj Bunin picked us up early and we went to the Belgian Army Staff HQ and were welcomed by the Director of Armour Col Vanderveen and had a briefing by him and, those of his staff and the factory reps, we would be working with. Mr Jaumotte from SABCA, the prime contractor, Maj Souchre, SO2 Log (equivalent), Maj Dion , OC Leopard Detachment at Belgian School of Armour at Leopoldsburg in western Belgium. Leopoldsburg is just 30 kilometres from the Dutch border town of Eindhoven the start line of operation Market Garden, 'A Bridge Too Far'.

Our visit itinerary looked very good with three weeks in Leopoldsburg, followed by a week in Rome then back to Belgium for a couple of days with the logisticians checking test equipment and manuals etc. in Ghent.

After the briefing we moved to Leopoldsburg which is set in nice flat country where the push bike reigns supreme although the camp reminded me of Puckapunyal and the town of Seymour. 'Hardly surprising!' everyone was very helpful and after we settled into our rooms and unpacked our gear we had lunch in the Officers' mess. One special thing we noted during the day was the Australian flag flying outside the HQ building. It was to remain there for our period at the school.' Nice touch!' Present for lunch were Maj Brunin, Maj Dion and another wing OC Maj Corchere plus seven Turkish officers who were also on a similar mission to us. After lunch it was don tank suits and into the tank for our first lesson. I learnt a safety lesson early when I was running through the setup systems and hit the Stabiliser-on Button by mistake. The gun breech flew up and nearly hit the turret roof and the turret traversed. I nearly needed a change of tank suit as it almost cut me in half. With the turret now under control we had a good rundown on all the turret fittings and, although the tank we were in had only the prototype turret, we learnt where all the boxes would go and what control's we had to understand and operate. Back in the school we sat down and went through all the lesson plans that would be used to teach us and latterly for all Australian training. After having dinner and a few drinks with some of the living-in young officers it was to bed perchance to dream.

The Belgians organise their meals differently to us. Lunch is a large formal meal with sherry and port before and a beer or two after, dinner conversely is usually a snack style cold collation of meat and cheese. We never did get to grips with it and used to go into town for our "usual evening meal' as well as the light supper at the mess. There was a nice little pub just outside the gates called the Windsor Hotel (?) so we would often have a beer there on way back.

Before I left Australia Jan had asked me to call in on a friend of hers Terri, and her husband Gordon and their children, Nicola and Andrew, who lived in St Albans in England. She had written to them and set it up so I just had to ring and confirm the dates. I rang and spoke to Terri and arranged to fly over on the Friday and spend some days with them coming back on the Sunday. But first back to the 'CHAR' French for tank (speaks it like a native!) and learn as much as we can.

To allow us to keep up appearances I asked for a driver to take me into town to do laundry and dry cleaning. No problem it should be ready either tomorrow Wednesday or maybe Friday, pretty casual system but 'when in Belgium!'

Maj Dion who was our prime point of contact was most helpful with arranging a car to take me to Brussels to catch the plane to the UK and Maj Brunin arranged one at his end to meet me on my return and take me back to Leopoldsburg.

27 April 1976 was our first day of theory on the TFCS and Andre Scheitler was our instructor and we soon became aware of his overarching knowledge of the system. Although it was a long day we felt we were learning and understanding all the interacting elements of the system and when it came time later to review the proposed Lesson plans for our schools it was simply a matter of 'Australianising' the somewhat literal translation from French to English. This task was to be an ongoing one but helped us as well. The need at times to ask questions, to allow us to ensure the lesson plans were readable by Australian students, meant we came to better understand it ourselves.

By now we were starting to be known around the camp and many people would come up and ask us questions about Australia. When I was telling one young soldier about White Pointers weighing a ton swimming off our shores I could tell he was thinking I was having a lend of him. Later, on Sunday 16 May, I went and saw the new blockbuster, 'Jaws' in Brussels. The young soldier might have believed me if he saw that movie or, perhaps, considered both the scriptwriter and I were smoking the same weed.

I had made friends with one chap who was a French-Canadian by birth but in the Belgian army. Naturally he spoke very good English and was a pleasure to talk to. Anyway for a change of pace he and four other young officers invited me to join them for a Chinese meal in Eindhoven. Now imagine the scene we have a French-Canadian born Belgian taking an Australian to Holland for a Chinese meal. As I said earlier the Dutch border town of Eindhoven was only some 30 kms away.

We spent the next day finalising our lessons on the theory of the computer and testing using reference cards and the self-test on the power supply. Then had a look at the hardware and played with the simulator Talafit which is a training module that allowed us to replicate the TFCS operation without having to sit in a tank.

This training module allowed us to quickly become familiar with the system and with Andre leading us through the analogue computer we soon became to appreciate how important the Talafit would be for training the Australian users, both the Armoured School students and the RTC students as well. Made a note to check the contract to ensure we are buying Talafit.

Lunch was with the Turkish and Canadian officers and later on we had a special dinner to farewell the Turks, Soup, Sole Normande, Tournedos, desert and nice wines. Back to room to find my laundry was back and only cost A $ 2.60, on mess bill.

It was a hands-on day today with a practical lesson on the fitting of the TFCS into a tank just back from the workshop. Stretch and I watched with interest how they went about the task as this was a basic task that all our tradesmen would have to do. First we had to find three of the four bolts that secure the sight; finally I looked in a bin, voila!

Naturally we observed and took notes while two experienced technicians completed the assembly. It took 45-55 minutes but their procedures seemed suspect to us but language barrier meant could not query them as to why. More than likely they slowed down so we could follow. Anyway our notes and practice should mean a time saving is possible.

With the tank back in working order we went out onto the range to run through the operating procedures. Noticed a fault but was able to rectify it, so must be learning. We took turns ranging all afternoon and soon had the knack of it. Dust effect seemed important giving false echoes so made a point to clarify that. Actually after a time we were able to determine any false echoes so it is simply a training and experience problem.

****

Friday 30 April 1976 – Sunday 2 May 1976 were set aside for the UK Visit. After breakfast I was driven to Brussels airport and paid Bf 6800 for a ticket on the 1010 hrs Sabena flight to Heathrow arriving about 45 minutes later and rang Gordon. He drove down to pick me up so I changed some money; US $50 became 29 Pound sterling. And had a cup of tea and waited.

Gordon arrived and we picked up Terri and went back to St Albans for lunch. I had picked up a few things: a bottle of Scotch for Gordon, Chanel No 5 for Terri, a doll for Nicola and a jeep for Andrew so it was like a Christmas lunch. Later Terri and I went shopping and had a fun time. Terri was a slim tiny girl who some years earlier had won what was the predecessor of Biggest Loser. She had lost well over half her body weight to win but had retained all her sense of humour and we giggled and laughed our way all around the Brent Cross shopping centre. There was a May Day sale so I was fortunate to pick up a pair of quality English leather shoes, (Church's) for 22 pounds marked down from 40 pounds. I still have them and wear them regularly today 40 years later. we bought a cake for dinner and wandered off having fun with the money side of it as I offered US dollar travellers' cheques, Pound Sterling, Belgian Francs and some leftover Dm so it was a case of 'let's make a deal'.

On Saturday morning Gordon had to go to work so Terri and the children showed me the old Roman town of St Albans. We walked miles looking at museums, roman town gates, the cathedral and the touristy shopping of George Street. After lunch, with Gordon and Andrew great Watford fans, watching the Fa Cup and Nicola at a party, Terri and I went shopping again.

I bought toys for Keiran, Jamie and David and a wallet or myself and then went shopping for groceries. They had already said they could seldom afford to go out to eat so Teri was keen to cook me a proper English roast dinner so we went out to the shops together. I was paying despite her objections and when it came to vegetables the prices were high by our standards with potatoes A $ 0.30 per pound. She said they seldom had them making do with crisps instead. I reminded her shop was on me as feeding another adult for a weekend was in my view too much of a strain on their budget. Terri complained and did not want me to, but I am bigger than her.

On Sunday Gordon drove me all over London and I took photos of Buckingham palace, Tower Bridge, Tower of London, 10 Downing Street, Horse Guards and Big Ben et al. Then home past Bank of England, Australia House, St Paul's, Fleet Street, Pall Mall and the Playboy club. 'Ticking all the boxes!' Back for lunch then a visit to the RAF museum at Hendon before Heathrow and Sabena Flt 610 to Brussels. My driver was waiting and we set off for Leopoldsburg, grabbed a take away Hambergerwurst and pommes frittes at a Fritur (a Belgian greasy spoon).

The next two days I sat down with Maj Dion doing a translation of the check list and confirming my understanding the system, not only what it will do but also how and why it does it. Now to get it typed up and photocopied. The tank was to be our place of work for the next few days and I found that a tank on the move is a very different beast to one stationary. We took turns in the Gunner's seat trying to maintain a sight picture with the TFCS and stabiliser on, while travelling at 20-25 kph across country. I did not make a success of my turn but found it easy to head-butt the sight, raising a large bump on my forehead that Sherpa Tensing Norgay would have trouble climbing. Later we worked with Andre doing fault finding and were successful in finding the faults he had introduced.

Final versions of the check lists were translated and filed away and we tried to set up a test scenario to duplicate a fault detected on the tank's trial in Sardinia where on a hot day there too many multiple echoes. We wanted to simulate the conditions with a small oil fire with another tank stirring up dust.

The test scenario was:

Do the sight/laser alignment test which gives a voltmeter reading proportional to the power of the laser return. Take the average of a number of tests and use this as a base, whilst aligned on the same target drive a vehicle in front of the target and take readings through the dust again still aligning on the same target light a small oil fire and repeat the reading through the smoke, and finally repeat the above but with both oil fire and dust.

The dust part was fine but the oil fire was too hard considering the hot day and wind blowing. So we had to rely on our previous conclusion of experience and training overcoming the problem.

LASER safety was the subject for Friday 7 May 1976 and we paid particular attention to this subject and made sure we understood and recorded the details as it was sure to be of prime concern. The morning after that was spent going through the translations of the TFCS description and signing off on it.

Being a nice day we walked into town and had a meal. Cordon Bleu and the largest meal I have ever seen, undaunted, by what looked like two regular serves on one plate I forced myself to finish it. 'The things I do for my country!'

On way back we stopped in the Windsor hotel for a beer before bed and met a couple, Emile a young guy and his wife, so we had two more with them. Now ready to go, I ordered a beer to drink whilst I wrote a letter, and Stretch left me to it. Commandant Resmet and his brother in law Commandant le Beau rolled in just then and that was that. They were 'crissed as pickets' and no way were they going to let me leave. Thankfully Mme Le beau came down to drag them both home so after a few more with her she drove us all home. The Le Beaus had a large MQ not far from my Room so invited me for a BBQ on the Sunday.

Note: Commandant, Cdt, is a rank between captain and Major.

****

With a weekend to fill we picked up ahire car and drove to Amsterdam in about three hours and booked into the Westania Hotel, bed and breakfast for US $12 per night. The owner has a sister in Bedford Park so had a chat then walked around the city and took a canal trip. Lunch was a chicken omelette followed by 'bonen' soup for dinner at an Israeli restaurant. Hot humid day and night and I slept on top of covers.

The return journey was a tour of Holland, once we made it out of the one-way streets of Amsterdam, we drove to the dike at Harlem/Zandvoort here we were on the beach, in beach weather, but continued on down through Den Haag, Rotterdam, Breda, and nearly to Antwerp. We had lunch at Westmalle where monks make Trappist Beer, a strong red beer.

The drive from Harlem to Den Haag through fields of Tulips all laid out in squares of vibrant colours.

We arrived back at 1700 hrs and strolled around the corner to the BBQ and drank Italian Red wine with a great group of people including a Catholic priest so the conversation was a mixture of French, Flemish, English and Latin. About 0100 hrs the priest produced two bottles of real Champagne and we enjoyed the experience of drinking fine wine in shirtsleeves in lovely company. Mme Le Beau paints for a hobby but to my eye they were way above the standard of a mere hobbyist. Finally at 0200 hrs we left and went to bed feeling much better for the friendships we had made.

This is to be our last day at Leopoldsburg and after only four and a half hours sleep, we staggered out to face what this day would bring. A good breakfast helped and we were soon in the tank doing complete first and second echelon, unit and field level tests. We can now discover the faults but just marked them for later repair and noted that the push to turn buttons were prone to stick and have no stop on them This session gave us the opportunity to correct and finalise the check list as we went along so now it is going to type.

Our next item to learn was how to correct the alignment of the sight reticule with the laser. To do this requires special tools so is definitely a second echelon or field repair.

After lunch we repeated the tests with Stretch in the Gunner's seat and me as Commander of the stationary tank. It worked well and was good fun!

The hot water had been off all day so we hoped it might be on when we get back to lines but no joy. Cold showers then we walked across to the mess for dinner and afterwards packed up for our move to Brussels.

Before we could go it was necessary to finish of some translations and get answers to some last minute queries. The Belgians took us to the NCO'S mess for drinks: the Chief Instructor, the Regimental Sergeant Major, Maj Dion, Stretch and I. It was very nice of them. Lunch at the Officers' Mess then I paid my mess bill Bf 2500, (A $50) before the drive to Brussels.

Our first task in Brussels was to find suitable in expensive accommodation as our advanced travel allowance was being used up faster than was hoped.

The first, recommended, cheap hotel did not have an en suite shower and toilet so looked elsewhere. The driver suggested an alternative tourist hotel the ABC Hotel. Each room had an en suite and twin beds, two chairs and a small closet. No TV, radio or phone but who cared it was cheap. Spoke to the man at the desk and he quoted" Bf 675 per night bed and breakfast." lapsing into my best Inspector Clouseau impersonation, "A NATO price monsieur?" I asked, "Certainment, Bf 400 per night bed and breakfast". That was about A $8 a night and much closer to what we were looking for.

That looked after the bed and the breakfast bit but laundry was still a problem. I visited four places and finally found one that did not have a 9-10 days wait. In our wanders we found we had ended up in the middle of the red-light district of Brussels. How did we come to this conclusion? It might have been the lovely lady standing on every corner, or it could have been the other ladies sitting in full view in the windows of the shops in the area?

Had another meal of escargots, deliberately this time but Stretch was not convinced and will wait to see how I surface next morning before committing himself.

The destination in the morning was the Ministry of Defence for a logistics meeting, present were: Maj Souquet, Program Section chief, Maj Roels and Cdt Plancke from the director of logistics, Cdt Le Coutere the systems management man, Maj Brunin and us. They had a presentation ready and went through it translating as necessary as they went as some of the presenters had little English. Our French and Flemish remained at its usual standard of course.

We broke for lunch where we were joined by Col Frank and Lt Col De Braix, the head and deputy head of Logistics. Lunch was a lovely three course meal of soup, Roast beef and desert with a nice St Emilion Red Wine.

After the splendid lunch it was time for more presentations before Cdt Le Coutere gave an explanation of the method used by the Belgian army to determine the spares needed to support X tanks for Y time. A fairly straight forward mathematical approach but only valid if the data in is accurate. We will seek user data on mean time between failures (MTBF) etc. but we will have to crank in very different distance and travel time and other local factors to the mean time to repair (MTTR). Our results will be different to the European expected values but I appreciated the method and made notes and asked many questions.

Thursday 13 May was a cloudy day and we were picked up at 0830hrs to go to the Arsenal at Rocourt, the Belgian army base repair facility.

Rocourt is just outside Liege so it was an opportunity to do a little sightseeing. Liege is a city along the Meuse River in Belgium's French-speaking Walloon region. Its old town is filled with landmarks dating to the medieval era, including the Romanesque Collegiate Church of St. Bartholomew, and well worth a visit. We arrived about 1000hrs and had coffee with the CO, Lt Col De Smet, Lt Col La Valle, Maj Souquet, and Maj Brunin. Afterwards we had a briefing on the Belgian Army system of repair and then saw the integration test equipment they use post base repair.

At night we walked to the Grand Place and listened to a band playing in a rotunda in the centre of the Place. Had an Italian meal for a change and on the way back to the hotel we stopped for a beer in a café. While we were enjoying our drink a drunk tried to get in. they locked him out, and us in, so there we stayed for an hour until the finally they called the cops and off he went for a night's rest.

The toilet arrangements made me smile. It happened at the Arsenal Officers' mess and again in the café. The arrangement is a urinal and a number of cubicles. I was standing using a urinal when a very fashionably dressed lady strolled past and entered the cubicle it was different but sensible once I got over the shock and thought about it.

Friday 14 May 1976 was to be my last day with the Belgian Army and what a day it turned out to be. It started at the Ministry of Defence where we sat down with SABCA reps and worked out our next few weeks. Everything seems to be sliding into place including a week with Selenia in Rome from 24-28 May.

Stretch took off for a week in the UK catching up with family and Maj Brunin and I went back to Leopoldsburg for the Feest van Kavaliers the Belgian Army birthday parade. It was a day of pomp and circumstance and beer and I thoroughly enjoyed it going to and fro we passed the Atomium, the Royal Palace and detoured through Leuven a lovely old town.

Saturday 15 May started late but I was able to pick up my laundry and although it was a beautiful day I sat in my room all day writing up my report. That night I had dinner with the Brunin family and they insisted I call Carol.

97

Simply dial 61 3 then number I think, it is different now but big deal in 1976. This was still during our estranged period so was not memorable.

The Brunin's and I took in the light show at the Grand Place and had a beer, Trappist, in a three storey café built in circa 1640, which overlooked the Grand Place. The Grand Place is extraordinary and a must see if anyone is lucky enough to visit Brussels. The columns and spires that are a feature of the buildings surrounding the main square are staggeringly beautiful. If you go down a lane to the left hand side of the square you come upon the Manneken Pis, "Little man Pee" in Dutch, a small (61 cm) bronze sculpture depicting a naked little boy urinating into a fountain's basin. It was put in place in 1618 or 1619.

When leaving the Manneken Pis and returning to the Grand Place you pass on your right hand side a bronze statue of a dolphin. Tradition has it that if you rub the nose of the dolphin you will one day return to Brussels. It worked, see later.

Sunday morning and up with the sparrows as what seemed like 9000 poms got up to take a bus somewhere. I wanted to sleep in but seeing I was up I thought I might as well write the odd letter then go for breakfast then do some tourist things. Having driven past the Atomium thought I would check it out. The Atomium was originally constructed for Expo 58, the 1958 Brussels World's Fair, and stands 102 m tall. I thought I could get there by tram but went in wrong direction so took a taxi. My impression was of a tired old relic of times past needing a face lift. The lift was exciting it went up the 100 metres in 15 seconds and came down the first 90 metres in about 9 seconds and braked in the last 10 metres. My stomach was somewhere in between. The renovation of the Atomium that I thought it needed began in March 2004; it was closed to the public in October, and remained closed until 18 February 2007. The renovations included replacing the faded aluminium sheets on the spheres with stainless steel. Its nine 18 m diameter stainless steel clad spheres are connected so that the whole forms the shape of a unit cell of an iron crystal magnified 165 billion times. It is now a museum.

Leaving there I walked back via the Royal Palace, Notre Dame de Laeken, along the canal to my hotel. With my report completed I took a taxi to the embassy to drop it off.

The embassy was closed but the watchman let me in to drop off the double-enveloped report. Deciding to walk back I found a magnificent clock at Mont des Arts not far from the Grand Place and had lunch about 1400 hrs where I could watch it do its thing. Found myself outside a cinema where 'Jaws' was about to start so spent an hour and a half or so being scared off swimming.

Found a stall selling mussel rolls so with a couple of them and a bottle of choc milk I retired to my suite at the ABC; the quiet end to a pleasant full day where I probably walked about ten kilometres.

The next two days were with SABCA and here we firmed up the test equipment required to support the TFCS at each level of repair. One thing that was obvious was that we would one-for-one replace all sensors. However, we were going for a full set of field repair plus two sets of base repair equipment for the electronics suite, and two sets of base repair equipment for the optics. It is impossible to do field repair on the optics due to the necessity of clean rooms to break the seal on the optical and laser systems. We have established facilities in Australia to do electronic module repair for other systems so it may be possible to co-locate this level of repair with those currently existing. The high theoretical MTBF appears valid and would suggest this approach as a reasonable one. All equipment exceeds

Military Standards requirements and given the commonality of component parts throughout the system and the test equipment supporting it , and the same or similar analogue circuitry, it will be easy for the technicians, as once they learn one circuit they know half the computer.

This all led to the following concept of maintenance:

- unit repair would be at assembly level replacement for all of TFCS,
- field repair would necessitate Nitrogen drying systems to replace electronic modules,
- no repair of optics at field level,
- whole of sight assembly returned to base facilities,
- base repair or industry to repair the modules and printed circuit boards, and
- Clean room 1 part per 5 million to do laser and optical alignment.

My next task was to request a table listing: components, part numbers, manufacturer, level of repair where needed, relevant assembly it is used in, general test equipment required and special tools, skills or test equipment needed. SABCA and Belgian army logistics were to provide this to Australia.

****

Our next stop over the weekend of 22-23 May was back at Bonn where we needed to debrief the TPLO and sort out our travel arrangements to Rome to discuss the laser range finder with Selenia the contractor. To make a change we were going by train, leaving Brussels from Gare de Nord on the Trans European Express (TEE) to Cologne, then by regional train to Bonn.

[Type text]

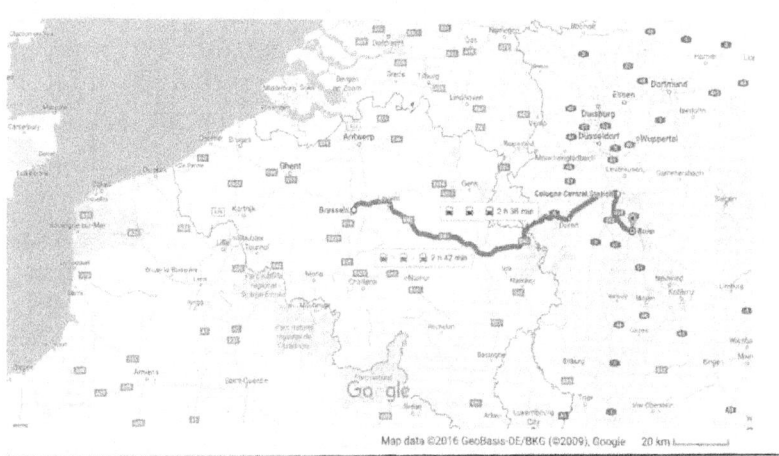

Go gle Maps   Brussels, Belgium to Bonn, Germany   6:17 AM - 8:55 AM (2 h 38 min)

The time from Brussels to Cologne would be about two hours, and then about 20 minutes to Bonn. The regional trains ran regularly so we did not have to make a booking, just walk across the Cologne station from where the TEE pulled in and jump aboard the regional. What on earth could go wrong?

The first part of our adventure was to sort out what we needed to pack to take to Rome and what we could leave in Bonn, so it was quite an exercise to work out both piles and then pack them accordingly. To add to the degree of difficulty, we had accumulated a substantial amount of notes, data and manuals over our time in Belgium so this had to be packed and stored somewhere but still be available to us later on. TPLO arranged this for us, it would be stored at the embassy, but we still had to bring it all with us from Brussels. In the end I had decided that we would be in civilian clothes in Rome so that allowed all the uniforms to be packed for storage in Bonn.

My luggage thus included a suitcase containing civilian clothing and accessories to cover me for just over a week, i.e. from Saturday and Sunday22 / 23 **May in** Bonn and Monday 24 May to Monday 31May in Rome, the cabin trunk that contained uniforms and the notes etc. from Belgium, a suit bag with my service dress uniform, and another suitcase with those items of clothing I could dispense with in Rome but would be needed again when we came back to Munich after we were finished with Selenia. Stretch had two cases and a suit bag as well so it was quite a pile on the pavement outside the hotel.

Giving ourselves plenty of time we booked a taxi and loading all our gear aboard we travelled the short distance to the Gare de NORD and arrived well and truly before the TEE was due to leave. I cannot remember whether we had to take the luggage on with us because we were going to leave the TEE system and go regional or we decided to because we were unsure if our luggage and ourselves would end up in the same place. Anyway, with a lot of lifting and pushing I stowed mine in a storage locker above my seat. Stretch was able to place his on the empty seat alongside him. A good thing as it turned out. This was my first train trip for a while and I loved the TEE. It was agreed by the two of us that any time we could we would opt to travel by train to enjoy the experience of fast smooth rail travel. Considering that we flew first class everywhere that is a huge compliment to European trains.

We were both tired out by the pressure and pace of the last few weeks and we were pleased to relax and watch the world flash by. We had a meal served, then, too soon, we were rolling into Cologne Central Station. We jumped up and disembarked trying to work out where we would find the regional tracks. I was hyper active looking everywhere until at last I spotted a departure time table board. It was around about then that I realised I was short a few things. I had left all my bags on the TEE in my rush to get off. Stretch had his bags but not me.

The TEE was enroute to Basel, Switzerland. "Dash it!" I thought and looked around for some assistance. There was an 'Information Counter, English Spoken' sign just up from where we were but there was a queue of about 20 or more waiting so I decided that I was more likely to start a riot than get help if I tried to jump especially not speaking the language. Next door there was another sign 'Casino' which was a lunch room for the railway workers. Reverting back to the old adage of 'faint hearts and fair maidens' I raced in and stood in the centre of the room and in a parade level voice asked for help. "Please I do not speak German and have a serious problem. Can someone, please Help?

A young clerical type came up and offered to assist me. He was nice enough, when told my predicament, to refrain from shaking his head in disbelief and sniggering, and led me back to the platform where the TEE had stopped. He said it would be easy to have the station master signal ahead and have the bags removed and sent back on the next possible train. It sounded like a plan, but would naturally mean a delay of some hours. However, once again, my old belief that *'God looks after Drunks and Fools'* was proven accurate. Sitting outside the station master's little glass walled office were all my bags. Apparently another passenger had noted my stupidity and had taken them off and handed them to the station master. It once again renewed my faith in humanity and at least two more were added to the Good Germans' side of the equation.

After all the turmoil we were pleased to have a quiet day or so in Bonn and relax ,compare notes with the TPLO and make sure the next phases of our evaluation was on schedule and all the administrative I's had been dotted and the T's crossed. Foremost in our minds, of these, were our travel allowance advances and reimbursement for ancillary expenses, e.g. taxis and purchase of extra luggage to accommodate the reams of stuff we were accumulating.

We had been booked to fly out to Rome after lunch Sunday and the embassy travel staff had been busy and all the liaison work had been done with Selenia and a contractor representative would meet us at the Leonardo Da Vinci airport and take us to our hotel. So as there was no problem with our travel arrangements, we were right for take-off for Rome. Little rest for the wickedish!

****

A week in Rome, and at the end of May, so the weather should be perfect, I was looking forward to our visit to a city I had loved from my reading of its architectural and artistic heritage. A number of must see items were on my dance card, work permitting of course. Seriously!

We flew Alitalia and were picked up at Leonardo da Vinci airport; Selenia had allocated a car and driver, Antonio, to us for the week and had set us up in a very nice hotel. The drive into Rome was of interest due in part to the number of young girls standing on the side of the road. They were prostitutes and this was definitely a niche market, pun intended, whereby clients would pull over pick them up and leave the road do the business then return the girl to the roadside and go on their way. 'Who said, when in Rome?' We declined the many offers we saw and made it safely to our hotel.

An English speaking Selenia engineer was there to meet us and after we checked in and had a drink in the bar he took us out for dinner. We guessed later that this restaurant was his favourite, or he had shares in it, because we went there nearly every night that week and Selenia were paying which was a bonus. It was my first opportunity to taste Palma ham and melon and this refreshing and flavoursome entrée became my first choice whenever I saw it. The Main course varied with lots of pasta and fish to tempt us.

We spent the week at Selenia learning as much as we could about the test and alignment procedures with emphasis on the clean room requirements. They confirmed what we had been told by SABCA so we asked to see the recommended test equipment and found them to be commercial in quality and not suitable for field deployment so that was a significant factor in our determination of our maintenance approach.

Our days were usually from early morning to just after lunch so we had time to roam Rome. We were well looked after regards our meals, breakfast was included in the Hotel tariff, lunch was at Selenia casino, and our English speaking engineer took us to dinner every night.

Antonio used to pick us up after breakfast each morning and be available to run us back to the Hotel after we had finished. During our last weekend Antonio and his car were still available. Antonio was a good guide and he recommended we visit the Tivoli gardens on the Saturday. We were a little sceptical, as we had seen gardens! This garden was a touch different because the Tivoli gardens is in a 16th-century **villa**, Villa d'Este,inTivoli, near **Rome**, famous for its terraced hillside **Italian Renaissance garden** and especially for its profusion of over 600 **fountains**. It is now an Italian state museum, and is listed as a **UNESCO world heritage site**. It did emphasise my disappointment with this modern Rome.

There must have been an election coming up because every building no matter if it was the average dwelling or the Villa d'Este at Tivoli and even the Coliseum were covered in graffiti. I was so disappointed.

[Type text]

After the Tivoli gardens Antonio suggested the Sistine Chapel. Now I had seen the movie so knew there was a magnificent ceiling so anticipated a quick once over and the usual tourist "Ohs and Ahs" then back on the street.

My Arthur Mee's encyclopaedia did not mention the chapel was part of the Vatican museum. So we were surprised when Antonio said it would take all afternoon and he would meet us at a café nearby.

Silly Italian, just because he lives here and is a guide why would he say such a thing. I had a better source of information. Naturally it took hours and I still would like to go back because we never gave the chapel itself the respect it is due. From one of the balconies we could look down into the Vatican City itself which was an unexpected bonus.

On the Saturday night, our last night we went for a walk and found Harry's Bar on Via Veneto. Bit of a problem there, we were running a tab and had had a few and possibly because we were so used to someone else paying we walked out without fixing the bill. Always meant to go back to sort it out but haven't managed it just yet. I haven't forgotten that I owe them a few (?) Lira, it is just a matter of finding the time to return and settle up.

Sunday was our last day as tourists so it was time to visit St Peter's Basilica and again my childhood memories of the statues of St Peter, outside the Basilica and the Pieta by Michelangelo came flooding back. Again we did not do this extraordinary place justice but we lucked into something else on our way out. It was noon and the pope came out to a balcony window to bestow a blessing on the crowd, including this little CBC Fremantle 'pagan'.

Later when we checked out of the hotel and tried to pay it was already paid for by the company. Considering our funding shortage at that time it was most welcome.

[Type text]

****

We flew to Munich on the Sunday afternoon and booked back into the Eden Wolff hotel. TPLO had made the bookings when we were in Bonn and transferred our 'left luggage' to our rooms.

This was to be a busy couple of weeks the first with Krauss-Maffei consolidating the warranty and support arrangements, then at Wegman in Kassel to evaluate the turret production and the location of the sight and control box mounts from both an operational and maintenance point of view.

Because this was a free period in Munich we decided to enjoy it and wandered around even though it was a touch cool and wet. When you are below ground level in the subway it is pleasant enough and in my travels looking for a paper I found a convenient laundry that was cheaper than the hotel. We dined out on a couple of Bratwurst and lagers; we were paying again, and returned to the Eden Wolff to prepare for our next and last week with K-M. The food in Munich is a bit disappointing after Brussels but I was determined to try before I left the Bavarian 'national' dish, Schweinshaxen, which is a roasted **ham hock.** The ham hock is the end of the pig's leg, just above the ankle and below the meaty ham portion. To me it was a pig's trotter and I found if greasy and uninteresting but maybe I went to the wrong restaurant. I intended to try it again somewhere else but failed to get the opportunity.

One good thing was that the price of clothes was cheaper than Bonn so I bought two shirts and some more socks. I will need another case before I return home.

Our last week at Krauss-Maffei started with us debriefing them on what we had seen and what we still needed to see to complete our familiarisation with the equipment. The gunner simulator is being purchased but we

107

had only seen mock-ups of the driver's trainer. This is all we will see before we go home so it is left to TPLO. Over the next couple of days we went through the warranty aspects with Dirk Hansen. The Maintenance Evaluation Team and Dirk are working well together but we still have some problems getting definitive answers to warranty implications of our repair concept. It is obvious that we must have a laser clean room but we must follow up on the possibility of feasibility of some repairs in a reasonable clean area and purging with nitrogen. For example, The Sight Electronic Box, Exit head, Binocular block and Instrument Light assembly could be removed and the sight sealed with plastic or paper with masking tape then a sealed transport unit opened and the module replaced. Very dangerous in a situation where dust exists but possible in Puckapunyal Wksp Company, armoured regiment workshop or RTC. Not advised for field workshops.

Thus I proposed and it was agreed that the only repair at field workshop level was to replace the complete Sight but it was possible to do all electronic module replacement. Higher level repair would be restricted to Puckapunyal Workshop Company, Armoured Regiment workshop, 4 Base Workshop, or RTC. I will urge Clean rooms for higher level repair to allay any warranty problems.

This agreement on a maintenance plan abled the MET to make a definite recommendation for the necessary test equipment to those charged with making the purchase. This left only some time with Herr Bremm of the Logistics Department and finalise the delivery schedule and sort out some variances we had noted in repair parts numbers. While we were doing our thing we ran into some Canadians doing the same thing for their tank purchase. We returned to Krauss-Maffei for lunch and more discussion with Dirk especially about the sight repair.

The TPLO was also involved with our discussions with Dirk but he had forgotten to bring our tickets for the next stage of our journey. Dinner was back at Ratskeller where Hitler, as said before, laid out his plans and policies for the formation of the Nazi party. Steak was tough but beer fine.

Thursday was an admin day with me picking up rail tickets to go to Wegman the turret manufacturer at Kassel which in those days was only just in American Zone and close to the East German border, and Stretch his tickets to Frankfurt-am-Main where he would assess the stabilisation system maintenance. The Cadillac-Gage system was identical to that used on the US M60 A1, Main Battle Tank so parts and maintenance support were assumed to be straight forward, and we simply had to verify this assumption.

With the tickets problem resolved we spent our last day with Kraus –Maffei. The first two turrets having been delivered by Wegman were in the factory for seating tests with the K-M made chassis. The first turret, AUS 17001, contained the classic control unit with German labels and the bore sight door would not allow the Lamp-Calc Test door to open or shut. AUS 17002 was fine. Both turrets would be returned to Wegman after seating tests in time for us to play with them.

I took an early mark and decided to go to the Olympia Complex a short tram ride away. It was interesting visiting a place that four years earlier had been the centre of the world's attention for all the wrong reasons. After climbing up onto a small hill and watching a number of enthusiastic flyers launching and remote control flying gliders I returned to the nearby U-Bahn station to return to the Hauptbahnhoff. This was my first time using the U-Bahn in Munich and was to be an experience. When boarding the train I looked for the ticket machine which in Bonn is just inside the carriage door and there wasn't one. "Oh, well," he thought, "will sort it out at exit gate." I noticed two scruffily dressed Germans, a man and a woman, giving a young English-speaking Irish au-pair a hard time, so moved over and asked her if she was alright. It turned out that she and I had the same problem we had not bought a ticket at the machine on the station before boarding. I tried to explain that it was a simple mistake of a stranger in a strange land and offered to pay then and there for both of us. This was my first experience with German officialdom and it made me realise that there were two distinct German people the nice ones and these lot. They must have enjoyed the 30's and early40's.

Not only could I not communicate with them I became aware that even if I could it was a lost cause. I tried to impress him with documentation, but, when I pulled out my Official, green, passport, he took it, glanced at it then put it in his pocket. This was going to escalate from fare dodging to grievous bodily harm if this cretin did not return my passport and in a hurry.

Luckily for both of us, a young student who spoke English came to my aid and acted as interpreter. By this time there was quite a crowd watching and my new friend told me the station master at the Hauptbahnhoff had been called and he would sort things out there. All five of us got off, the girl still crying, the two troll-like goons, my new friend and me. The station master spoke English and with my thanks the student left me and continued on his way. I never did get his name but he remains in my memory as on the Good German side. Once again I explained my reason for not having a ticket and offered again to pay for both the girl and myself. The station-master's first response was not promising as he said there was a Dm 60 fine due for each of us. I did not think this was fair and was prepared to move him further to the right of my dividing line of Good and Bad Germans when he finally grasped what I was saying about my passport, which was still in the male troll's possession.

He became even more officious, and sternly asked the troll for my passport. After looking at it, he asked what I was in their country for as an Official Australian Government representative. A short explanation, and the troll and trolless were told to leave and they obeyed, but not before giving me a dirty look, perhaps they were working on a commission basis. Anyway, it was all smiles and apologies, from the stationmaster. Once again I tried to pay but that was waved away.

Naturally I asked the Irish Au-pair if she wanted to come for a drink, but she had had enough of my company and, probably, Germany for that matter and took off.

We did not have anything to do before I went to Kassel, and Stretch to Frankfurt-am-Main, on Monday so this weekend it was tourism time.

Saturday was a real tourist destination. The Deutsches Museum in Munich, Germany is the world's largest science and technology museum, with approximately 1.5 million visitors per year and about 28,000 exhibited objects from 50 fields of science and technology.

We arrived mid-morning and by late afternoon having walked between 5-8 kilometres and thoroughly immersing ourselves in this treasure trove of technology we estimated we had probably taken in about one-third of what was on display. Returning to the Marien Platz we found a bar, not the most daunting of tasks in Munich, and had a pick me up or three, had dinner, then prepared to really be tourists tomorrow on what would be one of our last days in this city we had come to enjoy.

Having been known to have a beer on a warmish day we headed off to the Hofbräuhaus, a beer hall off the Marien Platz and near the Glockenspiel animated clock. Like all good tourists we were standing opposite at 1100 hrs and watched again when the 43 bells began to ring out and the

Glockenspiel recounted a royal wedding, a jousting tournament and a ritualistic dance - all events which have etched a mark on Munich's popular folklore. The show lasted about 15 minutes and concluded with the golden bird up the top emerging and chirping three times, telling us it was beer o'clock.

Inside the beer hall there was thumping oompah-pah music and 2 litre steins of beer being consumed with practiced skill. We felt comfortable! Finding a table with about half- a-dozen locals we soon joined in the merriment and by the time the next playing of the Glockenspiel at 1200 hrs we were steadily becoming fluent in beer-hall Deutsche, or Deutsche-lish, a much easier language than standard Deutsche, we found. I have no idea how long we stayed, or how much lager was partaken, but it was a great day. We were both up and feeling on top of the world the next day, Monday, as we prepared to split up and go our respective ways.

****

Our trains left at gentlemen's times, mine about 1200 hrs so it was a casual morning and a stroll across the road from the Eden Wolff to the Hauptbahnhoff. Having asked the question, this time we booked our luggage through and made sure we attached labels giving our current address, the Eden Wolff, and our arrival destination Hotel. If we lost our luggage it was hoped they would find their way home.

Kassel in 1976 was on the edge of the US Zones border with East Germany but no one had told me that at the time. The TPLO had mentioned in our first briefing to be conscious of the signs that designated the West/East German borders and steer clear because of the US Army Lance-Corporal had been snatched and dragged through the gates when he was taking photos near the border in Berlin, but at the time it did not sink in, as I had no idea that I was going to be so close to any borders.

Krauss-Maffei Wegman GmbH & Co. KG leads the European market for highly protected wheeled and tracked vehicles. The armed forces of more than 50 nations worldwide rely on tactical systems by KMW.
The trip to Kassel was made in lovely weather through very attractive country side in the very comfortable train which made European rail travel so special. I arrived about 1600 hrs and found that my luggage had not been loaded on this train but was coming (?). I left and went to the hotel and checked in. The room was more a small suite than a hotel room with a TV, telephone, double bed, lounge, armchair, footstool and an ensuite complete with a set of scales, I am not sure I will worry about them. The tariff for this comfortable accommodation was only DM 50 ($17) per night. Down stairs I found the bar and a barman that spoke English. (And I had been practicing my, "Ein Bier bitte.")

There was some music playing but the few patrons present were more reserved than their Bavarian cousins so my 'Guten arbend', smile and nod of the head went

[Type text]

unnoticed. Oh well, back to the barman and "Another one, thanks mate."

With little else to do I decided to have dinner in the restaurant at the hotel and tried the 'Polische' sausage. Later on I was starting to think that may have been a mistake. I had just finished eating, about 1830 hrs, and my luggage turned up. Tags obviously worked. After unpacking and another couple of silent beers I had an early night. 'Deutsche sprechen' TV was no incentive to stay up.

A vehicle picked me up and I met an English speaking senior engineer and because very few of the others at the factory spoke English I had an interpreter. We went back to the factory and I was introduced to the engineering staff over morning coffee and worked out my itinerary. There was no documentation available, in English, so it was a case of touring the factory and observing the construction of the turret on one line and then watching as they were lifted and placed on the structural fitout line where all the hard points were fitted to take the TFCS major subunits, sight, rangefinder, control boxes etc. The two turrets that had been in Munich, Aus 17001 and Aus 17002 were back on the factory floor and I observed, and had a play, while Aus 17002 underwent sight parallelogram testing. There were still only German labels on the classic system control units but I had already noted that earlier.

The next day I was in tank suit again and doing an ergonomic check. I simply used the experience I had with the TFCS fitted in the Belgian army leopards and tried to emulate the target acquisition and engagement actions in Aus 17002, which had TFCS subunit mock-ups welded to the turret. I ran through a complete drill engagement process and found a small but interesting problem. The Gunner's Central Control Unit #2 needed to be moved 3 cm to the rear and the bracket angled so that the gunner could see it. The wind sensor was sitting proud of the turret top, and any splaw-

114

footed digger could kick it off, let alone what might happen if it hit a tree branch. A wire globe protective cover was suggested to meet the requirement of recording air flow hence wind speed, yet reducing the likelihood of accidental damage. a canvas bag was also recommended to protect it from the environment when not in action.

The next few days were spent with the customer rep, and my interpreter, which was hard work as the interpreter seemed to be more anxious to impress me with his knowledge of English rather than acting as my communication link when things got confusing. In truth, as I realised later, it may have been the difficulty in correctly translating technical German into colloquial English. This problem became more obvious when I had to evaluate the manuals being offered with the turret. They were a literal English translation of the German but did not flow. For example: The tank commander's seat could be 'highered', not raised. This was just one of a large number and I spent the next day or so Anglicising, or more correctly, Australianising the literal English translation. Kept me off the streets!

My after work hours were usually spent in the monastic seclusion of my cell apart from a beer and dinner. "No one to talk to, all by myself." sounds like a lyric I could use as I was not misbehaving!

On the Wednesday night the English speaking engineer picked me up and we went for dinner then to his Shooting club. It was competition night and I was signed in as a guest and invited to participate. They used heavy barrel air rifles at a range of, I think, 15 meters at a paper target with ten bulls eyed rings.

We fired from standing supported, that is with your elbow or wrist leaning on a supporting frame, and standing unsupported where the weapon is fired from the shoulder but supported only by the strength of your arm, the left one in my case. There were two turns at each.

My first attempts were reasonable, but by the time I got to the last ten, firing unsupported, I was all over the shop, it was harder than it looked. Damned embarrassing! We celebrated with a number of Jägermeister then he dropped me back to the hotel. I hoped he did not have to go through an RTB.

The Saturday was to be something special. My Interpreter and a non-English speaking driver picked me up from the Hotel and we went touring. The plan was simple first we would go to the Hercules monument then for lunch.

The Hercules monument is an important landmark. Set in the Wilhelmshöhe Bergpark it dominates the surrounding area with a copper statue depicting the ancient Greek demigod **Hercules**. The statue is located at the top of a **pyramid**, which stands on top of an octagon all constructed at different times. Today 'Hercules' refers not only to the statue, but the whole monument, including the octagon and pyramid. The monument is the highest point in the Wilhelmshöhe Bergpark and there is a granite cataract with tumbling waterfalls and connecting runs from the top to the carpark level. The photo below comes from the internet.

After a fairly physical exploration of the whole area it was time to go and we climbed back in the car to go to lunch.

I was simply a passenger as the interpreter chatted to the driver as we drove through a wooded area. Suddenly I became focussed as we had just passed a sign that said we were just leaving the US zone. I was pretty short with both of them but was assured that we were not near the border itself. I sat back but was still a little concerned. About 5-10 minutes later we turned and climbed up a hill and turned into a beer garden with outdoor seating for meals. That seemed better then I looked out over the view to see kilometres of parallel wire fences with a bare area between them and towers every 500 metres or so. I was looking at the East German border.

The towers were manned and the bare areas I guessed were mined, and this was the last place on earth I should be. Given the furore that was raised when a US Army Lance Corporal was taken over the border, I hated to think what attention an Australian Army Major on Official business and with a high security clearance would attract if I was grabbed. I looked askance at the dickhead interpreter and wondered for a while if I had been set up but when I insisted very strongly that we return to the US Zone he grudgingly agreed, missed his free lunch. On our way back we drove for a time along the road about 50 metres on the Western side, of the first of the wire fences and then turned away. I was still concerned and angry until we passed another sign that said we were entering the US Zone. I was unsure if he was as stupid as he seemed to put me at risk, or whether I was just over reacting, but not fancying guessing incorrectly opted for discretion.

****

The next day, Sunday, I kept to myself in the hotel finishing off the documentation checks, ready to sit down and discuss the problems with the TPLWO Wolff Klimisch the following morning. We agreed that a proof reader was needed at all facilities to vet the documents supplied.

[Type text]

He and the TPLO were coming to pick me up so we could go and witness the proofing shoots of the first turret and gun assembly. To proof fire the turret and gun were mounted on a firing mound at a range near Kassel and three rounds were fired. The target was a cardboard square with a cross in the centre mounted on a fixed frame some short distance down range. The gun was bore sighted so the round would pierce the target at the centre of the cross. To prove the gun accuracy all three rounds needed to pierce the target at the same spot. If there was anything but one 105 mm hole in the target something was wrong. The test was successful and another tick was placed in the schedule of tests required before we could accept delivery of the tanks.

We returned to Wegman for lunch and were presented with an ashtray made from the brass shell cases fired in the proving, a nice gesture. I also souvenired the cardboard target used, got the proof firing range officer to sign and date it, and took it with me.

While the TPLO and TPLWO went on to other tasks I caught the train back to Munich and met up with Stretch and prepared to fly home on Wednesday. Our tickets were ready for us, Lufthansa from Munich to Frankfurt and Qantas back to Melbourne on Wednesday 23 June with an ETA in Melbourne mid-afternoon Friday 25 June.

First we had to pack all the documentation , reports, manuals, test results and other bits and pieces, including the target that I cut into a rectangle just large enough to fit the bottom of my cabin trunk. My final luggage included a cabin trunk, a kitbag, a large grey suitcase, a large beige suitcase, a small beige suitcase, a suit bag and a brief case. Stretch had nearly as much. Shifting this lot around was going to be fun, but not as much as we encountered. We had no problem with Lufthansa but when we fronted up to Qantas they had a coronary. The excess baggage they wanted to charge us was almost Dm 600, A $200, and no way was this little black duck going to agree to that. We were first in line and were not going anywhere until they saw sense. I showed them the magic green passport and our first class tickets and told them to give me something to sign and they would be paid by the queen but no joy. Ring the embassy I suggested but that was waved away, so still holding up the queue, I suggested calling down the duty manager. Eventually a chap turned up and we went through it all again, shades of the Munich train ticket problem. He had enough sense to recognise the problem so cut the Gordian knot and waved us through. We made our way to the first class lounge and had a shandy, refusing to think how we were going to get all this lot through customs at Melbourne. A problem for another day!

After a long but pleasant boozy flight we arrived at Tullamarine customs simply waved the lot through and we were home. Stretch's family had come to pick him up, and a major from Logistic Command had a vehicle to take me and all my goods and chattels to the Broadmeadows Officer s' Mess, where I had my room.

Chapter Nine – Overseas Training

On the way he sprang a surprise on me. Did I want to attend the 12 month Technical Staff Officers' course at the Royal Military College of Science in Shrivenham, UK, starting in December that year? He needed a quick answer as nominations were due. I think I managed to give him the quick answer he sought and I am sure he was not surprised when I said, "Yes, please."

RMCS traces its history to a school founded in Woolwich in 1772 to provide technical training for the military. In 1840 the Royal Artillery Institution was founded to train artillery officers "for the study of science and languages". It was established as a response to the technological advances brought on by the Industrial Revolution. During the 1880s, the institution expanded and a Commandant was appointed; it moved into Red Barracks, Woolwich becoming the Ordnance College before being renamed the Artillery College in 1918. In 1927 it became the Military College of Science.

At the start of the Second World War the College was moved from Woolwich, which was vulnerable to aerial bombing. It moved, initially to the artillery ranges at Lydd in Kent, then scattering to various locations until after the war, when the College was reconstituted and reopened at Beckett Hall in Shrivenham. In 1953, the college was granted its "Royal" title and became the Royal Military College of Science ('RMCS')

Since the early 70's Australian officers had been selected for training at RMCS to gain technical staffing expertise and fill the need for staff officers to be posted to the Materiel Branch of Army Headquarters, Canberra.

Mat branch was responsible for the staffing of the purchase of military Equipment, and its subsequent introduction into service, therefore, there was a pressing need for qualified staff officers given the need to re-equip and update the current Materiel establishment, post SVN.

There were two divisions of students at RMCS; Graduates were in Division One and non-graduates in Division Two. At the end of the year Division One students had to complete a thesis-like technical investigation and make a presentation to their peers, and Division Two had to write a report on a topic of military significance. The Division One investigation and presentation was done in teams of two or three depending on the complexity of the subject.

The intervening period from post o/s maintenance evaluation to o/s UK course was about six months but there was a lot to achieve in that time. Before I could think about RMCS I had to produce and publish the Tank Fire Control System Maintenance Evaluation Team's report to the Tank Project Leader so that the purchase system could proceed. Stretch and I worked in Logistic Command HQ in St Kilda Road and over a month or so massaged our notes and observations and the data and documentation provided by the contractors into a very thick document that was accepted and became part of the case for approval of the contract between the Australian Army and Krauss-Maffei.

****

Life had continued in my absence and I returned to find that the woman I had given up my family for was now living with another Army officer. I had been told of this while away so it was not a surprise when I got back, just confirmation.

When I was first told the situation it made me sad and angry but after a time I came to realise the extent of my stupidity and although, I felt a bit sorry for myself, I was well aware that it was a consequence of my own lack of character and decided that the time had come for a little self-evaluation and a bit more maturity.

Naturally I had been around to see Carol and the children and give over some gifts I had brought back with me but I did not press the point of reconciliation and continued to live in the Officers' Mess at Broadmeadows. A good friend and work mate, another Raeme Major Max Farrow however, could see that this hermetic existence was not doing me any good so insisted I move in with him. We enjoyed the title of the 'odd couple' for a time and he was a true mate and his sound thinking had a critical influence on me and helped me through this bad patch.

After a period of time like this, however, I realised that I was missing being a husband and father. When my posting to the UK was confirmed and as it was over twelve months, it became an accompanied posting, i.e. the family could go as well. This fact made me realise that firstly, I did not want to be 12 months overseas without my family, and secondly, this was a marvellous opportunity for them to enjoy overseas travel, something I could not have imagined being able to give them before this offer came up. Carol and I discussed it and I asked for a second chance. She very generously agreed and later on I moved back in.

Now began the pressure to tick all the boxes and make sure we had everything we needed before we set off. I had to hand my official passport in as I was on Student strength at RMCS and not official government business, so I was issued a standard blue passport in lieu. Carol had to get passport and as the children, were to be included on Carol's passport that was a priority. Naturally the report on the ME had to take precedence but while I concentrated on that we were looked after by the HQ Logistic command administrative staff who arranged for the passports and ticketing.

Max at this time wanted to go to Perth for time with his children, Colleen and Sean, after his marriage split up and we discussed the various ways he could achieve it over a brew one Tuesday morning. Being trained decision makers we raced past the obvious and decided that if we drove we could kill two birds with one stone, he would see his children and we could say goodbye to Win, Aunt Stella and Uncle Doug before going to the UK for a year. It was agreed that Max and I would share drive his Mercedes with Carol, Keiran, James and David as passengers. Some short time earlier the road across the Nullabor had been sealed all the way so it was doable.

The following week was clear so we could take leave. The plan was to leave after work Thursday drive non-stop and spend the following week in WA. We booked in to the Army holiday cabins on Rottnest from the Monday to Friday meaning we needed to be in Perth by Saturday to have our family days before catching the ferry to Rottnest Monday morning. Colleen and Sean were joining us so it was getting better and better all the time. Given that we would be back in Fremantle late Friday afternoon it was our plan to leave immediately and reverse the earlier trip and drive back no-stop with an ETA Melbourne sometime Sunday. It all seemed plausible at the time and on the way over it worked well. We drove non-stop to Eucla and had a sleep then continued on arriving in Perth Saturday evening. The week was a very good one with the kids having the time of their lives and Max and I while having a beer in the Quokka Arms Hotel met a fisherman and was invited to go out with him the next day as he pulled a shark line. Good option so off we went and had a good day at sea and took home a small school shark which we made into cutlets and BBQ'd them. It was our last night on Rottnest and made for a fitting end to our stay. Back on the mainland we packed the Mercedes again and started back East.

Everything was going well until, while I was driving, the motor overheated due to a split radiator hose but without catastrophic results. We limped into Horsham and stopped at a service station. I was distraught as I had not noticed the temp gauge going up and stopped, which might have reduced the damage. As it was, it was to be a new head for the good old Mercedes later on.

Given it was late Saturday night we were in a small pickle, but Max rang one of his girlfriends, and while we laid the kids down to bed in the service station she drove out and picked us all up leaving the Mercedes to be towed in later. Nearly a stress free voyage of adventure but 'missed it by that much!'

****

Meanwhile back in the UK posting section of our lives time was ticking down and there were still some hurdles to jump. We had decided to go via the USA to allow the kids to see Disneyland and everything appeared to be going well on that side of it when the wheels started coming off.

The first hiccough was a big one; the Brits had decided to start the Staff course in February and finish in December thus making it less than 12 month's long and that would mean it did not qualify as an accompanied posting. About the same time the bean counters decided to make overseas travel by officers, Economy Class not First Class. The latter was not as serious as the former. Luckily the powers that be realised that the RMCS class of 1977 might tell them to stick their course and as a lot of preparation had gone on, some had even sold houses in anticipation of buying in Canberra where we would all be posted for 1978-79, and had purchased cars overseas to take advantage of the reduced custom duties payable if you had been 12 months or more in country, the potential of claims for legal damages was on the agenda.

I have no idea who thought up the solution but it was so blindingly obvious and simple that it was probably a Corporal Clerk. It certainly would not have been a senior public servant. The answer was, "If they need to be on course for longer than twelve months and RMCS is now only eleven months, put them on another short course prior to the start of RMCS to make the total time over twelve months".

With that hurdle cleared it was time for some even better news. The bean counters, remember they also travelled overseas, amended the travel embargo on First Class travel to read, 'If an Officer was travelling overseas accompanied and had a child five years old or under, they would go First Class.' David was four and would turn five just before we were to return so the Masons were going over and back First class.

****

Although I had been a temporary Major since November 1975, one more little item needed to be addressed to allow me to be promoted to the substantive rank of Major in June 1978. I needed to complete the tactics course for major, Tac 3, at the Jungle Training Centre at Canungra.

My boss saw that I had finished my report and was only waiting and wasting time until we left so sent me on the next Tac 3 course, from 10 November to 8 December, which would finish about a week or so before we were scheduled to fly out. This suited us perfectly as it meant I could arrange the removal and storage of our furniture and effects from the MQ in Watsonia and give Carol and the children a holiday on the Gold Coast during my four weeks up in the hinterland. The MQs over in RMCS were fully finished so it was just clothes and the odd piece that we took with us, the rest of our goods and chattels were in storage for the year.

[Type text]

Carol and the children stayed in the Bilinga army holiday units and it was only a short drive down for me on the weekend.

My TAC 3 was a lot easier that the TAC 1 as I now had a couple of years' worth of field experience and was able to contribute sensibly to the discussions when our syndicate was preparing for the relevant TEWTs. When it came to the final TEWT I was actually able to lead the presentation instead of being a nodding head in the background as another syndicate member gave the syndicate's solution.

The last problem involved the deployment of a Task Force Maintenance Area (TFMA) and as this had been my bread and butter a couple of years back it was my time to step up.

Carol in the meantime moved out of the army holiday units and moved with the children to Brisbane where they stayed at a motel on Kangaroo Point.

Our diligent hard-working syndicate discussed the problem in depth over a dozen beers on the Friday night on the verandah bar at the mess and then decide we were ready so the next day we went to the coast arriving back on the Sunday night, ready for the presentation on the Monday. A fine RA INF officer Graeme (Moose) Dunlop did the defence aspects, and I did the rest, laying out all the TFMA units and marking out all the ancillary bits and pieces, access routes, vehicle parks storage areas etc. As I was one of the few, on the course or the staff for that matter, who had actually seen a TFMA on the ground, let alone completed a major exercise in one, at Singleton NSW in 1974, where I had been responsible for siting the workshop element of it, I basically 'cuffed' it and apart from a rough sketch had no notes to refer to. This was to be a problem later when we were basking in the glow of compliments from the DS on our presentation he asked for a copy of our notes and overlays so they could be used for future courses. What notes and overlays? We had not bothered with minor details like that. There were two solutions possible to sit down and write some up or ignore it and hope the DS forgot. Given that we were marching out in two days' time we ducked and hid and escaped without being sprung. It was not a problem for me I was leaving the country in a week to ten days' time and would be taking leave in Brisbane until our departure.

Carol and the children, Keiran nearly 10yo, James nearly 6yo and David just 4yo, had been luxuriating in a nice motel with a pool and they spent plenty of time in it until the clouds started gathering in the afternoon and David would panic and insist on going back to the room. While on the Coast there had been significant thunderstorms and he was not impressed. After I joined them it was time to sort out the final details prior to catching our flight. Our plan was to fly ex Brisbane and spend three nights in Los Angeles to allow the children to have a good time at Disneyland before flying direct to London. We were booked into a hotel in Anaheim just opposite the Disneyland gate. Qantas had been given our itinerary and had made the bookings for us so all we had to do is go to the main office in Brisbane and pick up the tickets. We took our brand new passports with us and went to Qantas and everything was ready for us.

I asked if they needed our passports for visas into USA but they assured us that, as we were transiting only with a final destination in the UK, we did not need a visa. With that sorted out we next went to RACQ and were issued our International driver's licences.

Given my last flight and the problems I had, of my own making, with the excess luggage I was looking forward to an incident and drama free trip with a nice family holiday in the middle. Silly Me!

When we arrived in Hawaii the immigration formalities were to be done there and then the manure hit the rotor blades. Not only did we have to have a visa but, as we did not have one, we had to leave USA within 24 hours. To say we were unimpressed is a masterful piece of understatement. It took the Qantas manager and the duty immigration officer to sort it out and allow us to enjoy our three days in LA, read Disneyland. Qantas were fined US$25k, and serve them right.

But their staggering ineptitude was still ongoing because when we arrived at Anaheim on the Sunday night we found that Disneyland did not open on Mondays, at this time of year (?). We had bought 3 day passes through Qantas and one day was useless.

Still we shrugged and copped it sweet, it was not all bad as we still had another couple of day to spend there before our flight to London, and we looked forward to two days of magic.

****

Although on the Monday Disneyland was closed, we had a tour of Pueblo de Los Angles and bought the odd touristy things and when we got back, the Anaheim Hotel/Motel where we were staying provided some entertainment during dinner.

We had arranged for the kids to enjoy room service in their room with Keiran as baby sitter whilst we enjoyed a pre-dinner drink and a quite meal in the restaurant. Everything started out fine with the kids meals served to the room about 1900 hrs and the TV on. We went down to the ground floor restaurant, sat down and had a whisky sour each, our drink de jour, and ordered a salad and prime rib for the main course. The salad duly arrived and was very nice and was enjoyed with another round of drinks. Given the big day the next day we ducked up to check the kids while

128

waiting for the main course and as it still had not arrived had another round of drinks. The restaurant was not full but there were a good number of patrons so we just sat and chatted and enjoyed our drinks, ignoring the passing of time. Sometime later there was a bit of activity with one of the American couples getting up and storming out complaining about the slow service and we heard some yelling from the kitchen area but not enough to stop us enjoying another drink. Eventually the maître d came and told us the reason for the delay, someone had sent back a meal and the chef had thrown a wobbly and slashed his wrists and they had to get the replacement chef to come in and take over. We were beyond caring and told him we were happy to wait, and checked the kids again. Finally we had our prime rib, and when it came to pay we were not charged for the meal, or for the entertainment. Luckily to get home we only had to stroll upstairs, for if we were driving any RBT would have been interesting reading.

After breakfast in the hotel we went across the road to Disneyland for our first full day of rides and fun. David was in front as we entered Main Street just as the Street parade was underway and there was Mickey Mouse leading. David ran up and grabbed his hand and hugged him and Mickey made a fuss of him.

The rest of us were delighted by it all and preparing to go on to some of the major attractions. David, however, wanted to go back to the hotel as he had seen Mickey and that was all he was interested in. He put on quite an act when we refused to take him back and it was a petulant grumpy 4 yo that was dragged around for the rest of the day. Even forty years later the others have not forgiven him for ruining Disneyland for them. I suggested we lose him and pick him up from the 'lost children's area' on the way out, but I was outvoted.

Wednesday came and we made it to Disneyland once more and this time saw a lot more before hurrying back to the hotel and preparing to leave the USA the next day.

The next day we caught a taxi to LAX early in the afternoon so as to be ready for our Pan Am flight to London. The flight was uneventful with all of us tired and with a night flight the kids dozed most of the way. Carol and I had a good meal with a nice wine and managed a little sleep before we landed at Heathrow and cleared customs etc. about 0900 hrs. We had been booked into the Embassy Hotel in Bayswater Road so took a taxi and arrived about 1000 hrs. Our rooms would not be ready until 1400 hrs so some jetlagged Australians with a pile of luggage sat in one of the lounges. I cannot remember if a restaurant was open or not but guess there must have been. I was running on about 1.5 cylinders by this time so it is all a blur. I did change a traveller's cheque to give us some cash but that was about it.

Finally we made it to the rooms, two connected rooms on the first floor and after ordering room service for dinner we all crashed quite early. About 2330hrs there was an excited Jamie jumping on our bed yelling, "It's snowing, it's snowing!" so we all sat up to share the excitement of the kid's first snow fall, early morning 19 Dec 1976.

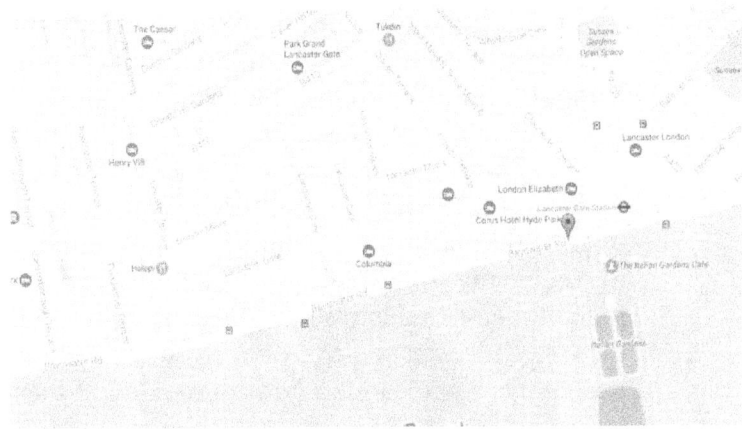

Bayswater Road London

The Hotel was supposed to be 4 stars but without in-room coffee-making and room service coffee at 2 pound sterling a cup it did not rate with me. Sunday was a lay day with a walk around to orient ourselves and apart from breakfast we ate out finding our way around on the hit and miss principle of walking turning at a corner walking turning another corner and so on until we were back. Naturally we crossed the road to Hyde Park but the weather was too cool to linger, especially when we only had Queensland relevant clothing. I found a handy tube station and this was to be important the next day, Monday 20 Dec 1976, as I had to report in to Australian Army Staff (London).

Head of AAS (L) was Colonel, later General, John Grey and he had a fearsome reputation so I decided I needed to have a haircut before fronting him. The train took me to the Strand, near Aust House and I found a barber and without looking at prices or whatever had a haircut that cost 15 guineas. This was the only time I was charged in guineas and wonder if I was conned. It was a damned expensive haircut but that may have been central London prices. I was not to know so just paid. Every other haircut in the UK was in a provincial area or at the RMCS on site barber so I have nothing to compare it with.

131

My interview was short and sweet and confirmed the itinerary and programme we had been given in Australia before leaving. We were to move down to RMCS late December and move into a MQ, settle in and then attend a short 4-5 week course, prior to the beginning of the Technical Staff Officers' course in late February. I was told to go to the Bank of NSW branch nearby and open accounts to receive my pay and allowances, I was reminded to use my rank of major as this would smooth all official procedures in the UK. This was to prove correct. And although it felt like posing, we soon learnt to go with the flow. When in Rome!

With the courtesy call out of the way and the important aspect of money under control we went shopping for climate suitable clothing finding a handy Marks and Spencer that was able to satisfy our needs. David opted for a lined yellow with blue woollen lining Paddington Bear coat that he wore for years before it was passed on to a younger cousin.

The strength and durability of the coat was to be tested later by an Oxfordshire canal. Keiran and James had overcoats and beanies and we were much better equipped. We spent the next few days sightseeing with visits to the Tower of London and Westminster Abbey as well as checking out Harrods. By this time we were sick of the Embassy Hotel and, as we had to move anyway in a week or so, I checked us out, and with AAS (L) help, booked us into a suite at the Post House hotel in Swindon which is only a short drive from RMCS.

One of the things we had arranged prior to leaving Australia was the matter of buying a car. There were basically two options, make use of the stamp duty concession afforded Australians who had worked or lived overseas for over twelve months and pay only 48% duty on the car purchased overseas and used for longer than twelve months, or buy a second hand car to use in the UK and sell it before you went home, the backpackers' approach.

Anyone who visited Aust House in London at that time saw a fleet of camper vans parked outside with "For sale" signs on them. I thought of buying an Aust design rules model, Peugeot and even paid a deposit but then decided it was an extravagance and the backpacker option was a more economically sound proposition. I contacted a Raeme friend, Jack Smiley, who was doing the course before us and he offered me his Automatic Rover 3500 V8, for A$1500. It was a good deal for both of us as the sale of V8 cars that had to have 5 star fuels at over 2 pound 50p sterling a gallon was more than the average Brit could afford and he knew selling to me was easier than trying to sell locally. Similarly, I knew I was getting a vehicle that had been looked after. Twelve months later it was our turn and I sold it to another Raeme mate, John Wilson, who was at WAIT with me, and was to do the course after us, for A$1200.

Jack met us at the Post House and we fixed up the car formalities and later the next day I went into Swindon and completed the transfer of registration. It was now time to have a look around and we drove out to Shrivenham, the nearest village to RMCS and found the local bank and arranged to have our account s transferred. The college was a short distance along the Farringdon Road part of the A42 that goes to Oxford so we headed off for a quick sticky-beak. The MQ we were allocated was in Wellington Square, Watchfield which was across the Farringdon Road from the college See map below. The college is on the right.

We called in, on the off chance to see the couple we were taking over from, Col and Marlene Dobie. They were our age and had two boys not much different in age to our children and we were later to become the closest of friends, but at that time we had not met.

We knocked at the open front door of the MQ and could see inside evidence of packing up and preparing to await removal so we were ready to just say hello and then

133

leave. As we were about to wander off a large man in overalls came down the stairs. "Hello Col," I said, and started to introduce myself when the overall clad man began to talk in a foreign language. I thought I had gone to the wrong quarter and this guy was a foreign student when he was followed down the stairs by a lovely very blonde lady who in an Australian accent introduced me to Arthur, who was helping out. Far from being the Australian Army officer I was to replace in the MQ, Arthur was a retarded man, about 30 odd who lived nearby and was used by the college as a handyman and general helper when minor jobs were required. He spoke in a broad Wiltshire accent and even by the end of the year I had difficulty understanding him. Col was out so, rather than take up any more of Marlene's packing time, we agreed to meet for drinks at the Post House later that evening. We were to spend a couple of weeks, including Christmas, in the Post house waiting for the MQ to become available and the children actually started school at Watchfield primary school while we were still in the hotel.

Chapter Ten– RMCS, Shrivenham UK

The Royal Military College of Science is a tertiary college providing science degrees to UK and some foreign subalterns, as well as the Technical Staff Courses for UK and foreign students that we were attending. Like most military colleges of its size it fielded sporting teams that competed in both military and civilian competitions. The Firsts rugby, cricket, tennis and hockey teams played at regional level and blues were awarded if you played at that level. In the previous years there had been some exceptional Australian 'blues' winners but our group had to settle for playing in the lower 'owls' competitions. You were awarded an 'owls' tie if you played in these college teams. About 5-6 of us played rugby and a few others played tennis.

Aerial photo of RMCS

[Type text]

Married Quarter

Our UK MQ was the house on the right where the first white car is parked and we were very pleased when we finally moved in and began to settle in to the local scene. The MQ was a fully furnished 4 bedroom brick house with central heating and a lockable garage on the left hand side. It was very comfortable and only a walk or pushbike ride from where we worked.

****

We bought two push bikes from a Canadian couple who were on the way home and they were very useful. One of the little features we had not struck before was called a cool box, a box cut into the kitchen back wall that was open to the elements.

We stored our dairy products in it and given the ambient temperature at the time we moved in it was not necessary initially to use a fridge. Although we had the cool box we did buy the Dobie's fridge as they were not able to take it with them and we bought a washing machine and our first dishwasher, a luxury indeed in 1977.

This is a good spot to explain the difference between the pay rates of the Brit Army and ours. In a word the UK government were parsimonious, at best, with the pay rates of all the services, military, Naval and Air plus the fire brigades and police, the fire brigade went on strike during our year in the UK. Unless the Army or other services were deployed overseas they were POOR. It needed them to be on allowances to live what was, to us, a normal life. For example, at RMCS an Australian Army major, on allowances, earned as much as the Major General who was the CO of the College. As well as the dozen or so Australians there were a handful of Canadians and one US air force officer on the course all of us on allowances and faced with the same delicate situation.

Part of our briefing was a stressing of this disparity; warning us not to flaunt our relative wealth. We were advised to make hospitality low key and of the "at home" style rather than elaborate lobster and oyster style multi course extravaganzas, as the Brits could not reciprocate and it would prove embarrassing.

We formed what was called the Canaust club and worked out a way of meeting our own social desires whilst still involving the Brits as much as we could and with as little ostentatiousness as we could. We made a programme of social events that began with a regular Friday night drinks and dinner moving each week to a different location. The evening would begin with a happy hour at the Shrivenham Arms Hotel in the High Street. UK had strange, to us, licencing hours where they would close at 1400 hrs and reopen at 1600 hrs. We used to meet about 1630 hrs and if anyone arrived earlier than 1600 hrs we would park in the back yard and enter through the back door. Given the drinking habit of the locals, who would have a pint and sit on it for an hour, and we would have three or four in the same time we were very popular.

The "at home" is a Brit convention where you leave your cards with the host who advises left their cards when they are "at home" to receive guests. A usual "at home" starts about 1730 hrs and all are gone by 1900hrs at the very latest. There are drinks and nibbles provided and it is intended as a 'meet-and-greet' type of function.

The Canausters remoulded this simple little social get together by having wet and dry food servings available to supplement the nibbles. They were not formally presented, just placed on a side table about 1830 hrs and guests were offered the opportunity to help themselves. It was usually presented in a casual way, "Carol had made a few things if you would like some." This kept the majority past the usual dispersal hour but as, we hoped, it was assumed to be a strange colonial concept it would not then be seen as imposing any obligation on our guests. The drinks were similarly available including, when available, Australian origin, Fosters and Australian wine sent down from Aust house and offered on the have-you-tried-this basis.

We usually avoided having a display of spirits because that would have been too much, the odd pink gin being the exception, if requested. We usually did this about every month or six weeks and managed to invite most of our UK classmates over the year; some became special friends and became regulars.

Another great disparity between the Brits and us was the meat we consumed for our family meals, they were lucky to have mincemeat and we kept eating the type and quantity of meat we were used to. I bought a freezer and using a contact passed down by Jack Smiley went down Farringdon road to Mushroom Farm where the farmer killed and hung his own beef and Lamb. The farmer was used to us and knew the type of things we consumed so when I ordered a hind quarter of beef and a side of lamb, he knew the types of cuts etc. we liked and we would pick up the order later after leaving him hanging and aging and butchering and packing them. Included would be some ancillary packets of sausages and offal if we asked for them.

Our biggest problem was getting the fish we were used to. Cod and ling and hake were freely available but Schnapper or 'heavens above' flake were very much a no show. A number of times I went into the Fishmongers in Farringdon and tried to convince him to get me a small school shark or equivalent but he probably thought, and perhaps does to this day, that I was pulling his leg as I never got one. They were caught in Cornwall, but to a fishmonger in the centre of the UK, that was a very long way away.

All the children got bicycles and it was David's first and had trainer wheels. "I do not want them." was the cry of the 'grown up 4 yo'. James and Keiran did not have any so that was it as far as he was concerned. I took them off and away he went with me holding the seat to steady him. That was not good enough, he had to fly solo so he launched off to land on his nose, blood everywhere, but undaunted he kept trying and despite more spills finally got the knack and there was no stopping him. I sent him a copy of the MQ photo this week, 41 years later, and his first comment was, "I remember riding my bike up and down that street."

A few nights after we moved in we had to evacuate and assemble on the oval at the end of Wellington Square when there was an IRA bomb scare. There was a flurry of activity over this time and although we were subject to a number of scares luckily we were not attacked.

One still lives in my memory. It was just after the mid-year break and about 1400 hrs when we were evacuated from the classrooms to assemble on the car parks. I was standing quietly dreaming of, probably, a beer after work and I was one of a number called out to assist in the search of a physics

[Type text]

lab in a science building to see if there was anything there
that should not be; a bomb for example.

I was partnered by one of the civilian instructors who
taught in that lab and knew it well. I could not stop thinking
how lucky I was. Of all the officers there, including Royal
Australian Engineers and Royal Engineers or Royal
Australian Ordnance Corps and Royal Ordnance Corps all
who are trained in explosives etc., they chose a Raeme
Electronic engineer?

We searched and searched and it was a relief to find all
the equipment in the lab cabinets and on the benches were
those items that should have been there. After about what
seemed an hour I started breathing again as we finished our
search and dispersed.

In early January the RMCS course was still a month
away so the Australian group were sent on a Chemical
Defence Science course that covered the problems involved
in fighting a war in which Chemical agents were deployed.
The content and even the location of some aspects of the
course involved some security problems so we had to have
been especially cleared prior to taking the course.

There was a significant amount of theory, almost all,
secret or above, and some practical where we used the UK
issue NBC protective dress, or 'noddy' suits as they are
called, in gas filled rooms and while performing normal
infantry functions marching, running, stripping and
assembling weapons and firing range practices with a
number of weapons. It was all new and interesting and we
had a ball.

Finally it was time to begin our course and we were
briefed by the resident Australian DS who was to help us
over our year. There were a number of traditions that we had
to be introduced to and prepare for, the formal dining in
nights and balls that would be held over the year, the sports

we could and should be involved in, mainly for me rugby and golf, and most importantly the Australian Day where we entertained all the students, staff and their partners to a lunch. A senior student, an RAAC officer Kevin Fletcher, was appointed and a couple of others of us formed a committee to assist him in organising it all. I was chosen as the Canaust rep and with my Canadian counterpart set about preparing a programme of social events which include two Canaust Dining in Nights to be held at the Shrivenham Arms, one in each semester. Our predecessors had arranged a golf long weekend visiting the Open Championship venues, so I undertook the task of duplicating this as well.

Chapter Eleven – Golf in Scotland

1977 was the Queen's Silver Anniversary year and there was a special bank holiday promulgated making the usual long weekend a day longer from 4-7 June inclusive, so these were our target dates for our golf tour. To make things easy I was able to access the previous year's file and used this as a template for my approach. The first thing was to select the venues and then build a program around them. The first venue was easy, St Andrews, but the others were up for grabs. In 1976 the guys stayed in the town of Kinross 400 miles and just under 7 hours away, just over the border and close to Edinburgh, but only 45 minutes' drive from St Andrews so we decided to do the same and select the other venues a similar short drive away from this base. A check of the maps showed that Carnoustie (1 hour), Muirfield (1 hour) and Gleneagles (30 minutes) were relatively close and we opted for them as our choices. Our plan was to play one course each day and that necessitated us taking an extra day 8 June as a stand-down day but the college would not approve it so we had to either only play three courses , or two courses on one day. We chose the latter and played Gleneagles King Course on 4 June, then The Company of Edinburgh golfers' course at Muirfield on 5 June, and elected to play Carnoustie in the morning and then drive to St Andrews (40 minutes) and finish on a high at the home of golf, the old course on 6 June.

To accomplish this I wrote to the secretaries of each club and requested a visit that included a round of golf and in the case of Gleneagles and Muirfield use of the club house and lunch. All acknowledged and approved our request so we were ready to go.

Golf at RMCS was played at two levels with the less fanatical playing on the College course and others being granted honorary membership of Frilford Heath Golf Club a very exclusive and significant Oxfordshire golf club. Those of us who chose Frilford Heath were made most welcome although our honorary membership limited us to play on designated days only, i.e. not weekends, it did not worry anyone.

We decided to leave after lunch Friday 3 June and drive to Kinross and book in ready to go the next morning. Our group was composed of about half a dozen Australians and a couple of Canadians and we drove up in a campervan driven by Kevin Fletcher and two cars. The Brits wanted to know where we were overnighting on our drive to Kinross not believing we were going to drive it in one go. With it being midsummer the days in Scotland were very long indeed and I can remember walking around in Kinross about midnight and it still being light.

That Saturday Scotland beat England in football at Wembley and the Scots in the crowd invaded the pitch after the final whistle and dug turfs of grass out to take home as souvenirs. In the Kinross pubs there was a lot of celebration and I remember entering one where there was a group playing and as soon as they heard me order a beer they broke into 'Waltzing Matilda"; a good guess or perhaps a hint of accent.

****

Not to be left out Carol and another Australian wife Noela Stamford, took off for a weekend in Paris a little earlier in May while I looked after the kids; a fair exchange.

Noela's husband Mike was a very special sort of man with an ability to make fun of any situation without regard for any sense of decorum.

143

[Type text]

For some reason he thought he could sing and would launch into his version of the only song he knew any time he had a couple of beers. I used to encourage him wickedly and on one memorable night stood up in a pub and announced him as a special act all the way from Australia traveling the UK on tour. He was overjoyed and launched into his song to the dismay of the locals who were too polite to throw things but were not amused as he murdered the song while making all the facial expressions and arm movements expected of a real entertainer. Noela used to try to hide and disown him but that only made him worse. Carol used to be annoyed with me for egging him on but it seemed good fun at the time.

On our first college dining-in night we stayed late to close the bar. When we were finally booted out Mike invited us back for a drink, obviously what we needed.

There were about five of us and Mike decided to give us breakfast making scrambled eggs and bacon and serving them on his best dinner setting with their best silver. We found out later that he had used the setting that Noela had, earlier that night, laid for a dinner in their dining room that they were hosting the next night. I think we all copped a serve from our wives over that one.

One Friday night Carol and I had just arrived at the Shrivenham Arms when we were called out and had to go home. David and James who were supposed to be home with Keiran as babysitter had gone out and crossed the road to the college gate and when coming back David was dared to run across the road and did so only to be hit by a car and taken to hospital. The car was a brand-new BMW driven by another Australian, Miles Nelson. When we analysed it later we could thank the fact that it was the BMW. If it had been a local vehicle it would never have stopped but the BMW slowed so fast that there was no injury to David, just a fright and a wakeup call on road safety. The other boys present,

James included, were similarly chastened for I think they had dared young David to try to run.

After a time we could take him home from the hospital and it was a very relieved family that arrived back at the MQ that night.

Our children were able to gain a lot from the time in the UK with Keiran and James attending the local primary school and David going to a preschool at the college itself. Carol took James to start at Watchfield and when the teacher asked his name she started to say Jamie and the 7 yo piped up and said my name is James. The teacher looked at him and said, "Fine. We will call you James." And from then until now he has been James. He had been watching TV and the 'Bionic Woman' or 'Six Million Dollar Woman' was Jamie Sommers and he was not going to be known by a girl's name. I remain the only one who calls him Jamie, but he accepts that.

Chapter Twelve – The Technical Staff Course

The Course was intended to give us the skills to investigate a piece of Military equipment and write a technical report on it listing its assets and limitations and making a series of conclusions and a recommendation on its value to the Army. Mostly we were offered two similar pieces of equipment and you had to make a comparison between them and recommend which one to purchase. For example, we had to compare the Swedish S1 Turretless tank with the UK Chieftain which was turreted.

The main paper we had to write was a thesis on a subject selected by the college and foisted upon us. It was done in pairs or threes depending on the depth of investigation required. You may gather that I would have preferred to have a choice and nominally we did.

There was a list of subjects and we could ask for them. I wanted to do a statistical study of WW2 determining the efficacy of different weapon systems, small arms, mortar, artillery, and armour by analysing killed in action and wounded in action numbers allocated to each system. It would require access to war diaries and casualty lists but seemed an interesting subject. Not to be I am afraid, a team of Brits got that one.

Another Raeme guy, Kerry Tunbridge, and I were given an 'Investigation into the effects on a Soldier's Performance from the need to Wear NBC gear in Hot /Dry and Hot/Wet Environments' this was due I suppose to both of us having done the Chemical Defence Science course in January. Now as a subject to investigate and report on it did not appear at odds with the probable need of Australian Soldiers to operate in hot and wet or hot and dry conditions and if they were ever required to employ NBC protective gear it would be nice to have a report that addressed the likely ramifications of said employment. However, I was wary of some minor caveat that may alter the apparent easy and obvious and cloud it in a veil of misty and unexpected problems. I had by now seen how most things that appeared on the surface to be straightforward, such as returning to Australia with some excess luggage or booking a few days in Disneyland could turn to 'merde', this simple subject carried with it its own little caveat. Those of us who read spy thrillers or watch the movies will remember *For Your Eyes Only* (1981) the twelfth spy film in the *James Bond* series, and the fifth to star Roger Moore as the fictional MI6 agent James Bond. The title comes from a security clearance level that precludes anyone but the addressee from reading the classified document. In our case the security classification placed on our report and any notes taken was 'UK EYES ONLY' that meant that no one but UK citizens, military or civilian, could access and read the report or our notes. Yes, that meant that, although it was investigated by and written by Australians, it was not able to be read by other Australians including those in AHQ Operations branch responsible for planning operations in hot/wet or hot/dry conditions who may be thinking of adding an NBC element to the exercise. There is a word for this but it is not used in polite company. Still we gave it a go but I can say without breaching any security that it was not given the attention it deserved and I have always been a little ashamed of the result but with no purpose obvious to us it was not surprising it fell short of our best. We received what used to be disparagingly called a Commonwealth 'C' and put it behind us

To give us an opportunity to see a variety of equipment and how they were deployed we went on a visit to the UK army schools Royal Artillery at Larkkill and the Royal Armoured Corps at Bovington , saw a deployed signals regiment in Germany and a live firing exercise at Salisbury.

****

Our other visits included a trip to British Aerospace in Bristol where we inspected the assembly and fit out of the fuselage and wings of the Concorde. Once completed in Bristol they were loaded into a specially set up transporter and taken to France to Aerospatiale where they were assembled and had the Engines and Avionics fitted and the completed aircraft flight tested. The highlight of our visit was the opportunity to inspect the interior layout of a demo aircraft fuselage, very narrow. I always wanted to fly in one but not to be. They used to do pilot training with bump and go landings and take-offs from RAF Brize Norton that was not far from us. Our MQ was right on the flight path so we were regularly treated to a close fly past as they went over us with the drop-nose still down.

To add to our knowledge and widen our experience we were able to look at another couple of defence industry establishments, RAF Fairbairn, AWE Aldermaston and the Royal Ordnance Factory at Nottingham, each of which gave us a special opportunity that we would never have been able to achieve in Australia.

The RAF Fairbairn is a research establishment with both aeronautical and environmental testing facilities and I was to return there later on to do our own testing by wearing a 'Noddie' suit in various climatic chambers and recording our core temperature and ability to perform various tasks as the climate varied. Another test was to assess fluid loss and this luckily was conducted by qualified personnel and we were just the guinea pigs. We had access to reports from actual deployments at the Ministry of Defence (MOD) in London and made a number of train trips from Swindon to London to make our research.

We revisited some of the places we had visited doing our Chemical Defence Science course and were accorded all the help we could ask for. At least the travel allowance was adequate to facilitate the research, even if the motivation was lacking.

The Atomic Weapons Establishment (AWE), which develops, maintains, and disposes of the UK's nuclear weaponry is in the parish of Aldermaston. Built on the site of the former RAF Aldermaston, the plant has been the destination of numerous Aldermaston Marches and it was a special day when we drove through the gates and past the sites where the marchers had occupied during their demonstrations against nuclear weapons. We had watched it on TV but it was now all there just outside the bus window.

****

ROF Nottingham was of particular interest to me as they were major producers of large calibre weapons including the Australian Leopard tank's main gun the Royal Ordnance L7A1, 105 mm tank ordnance, which was for a long time one of the most important products, of the ROF.

The specifications of the L7 A1 are:

Calibre: 105 mm

Barrel length: 52 calibres

Weight: 1,282 kg

149

[Type text]

Length: 5.89 m

One of the most interesting things we saw was the way the L7A1 was made. As shown above the gun weighs approximately one ton and is nearly 6 metres long so to see this large piece of round steel being lifted off supporting frames from the horizontal and swung into the vertical by overhead crane was impressive enough but then to watch the crane manoeuvre it into a vertical furnace where it hung perpendicular to the ground and was heated until red hot, was quite a sight. This, however, was only the start of the show. After a certain period controlled by the master Armourer it was removed and laid horizontal on a steel frame and the master armourer, by eye, used a crane borne hammer, a solid block of steel to drop down and hammer out any bends in the steel slug and make sure it was straight. It was fascinating watching this steel hammer being positioned along the line by the Armourer using an overhead crane and then to drop it and make necessary coarse adjustment until he was happy that it could go on to the final alignment boring, rifling and machining shops. the armourer was nearing retirement age and had two 'apprentices' working with him and learning the skill of this artisan.

Our trips were usually by bus and a group of us would grab the back seat and set up a table and played bridge as we motored along; very much a mobile Gentleman's club. "Play 500? How low brow. Bridge sir, if you do not mind."

****

The other big plus of living in Wiltshire was that we were very close to the centre of Celtic England.

Stonehenge: Wiltshire's most famous historic site built 2500BC

Avebury: Stone circle is thought to date back 4,000 years

Westbury White Horse: Cut in the hillside in the 16th Century

Within easy drive was Stonehenge and a drive down there and the chance to walk around and climb on and over the stones was a never to be repeated experience. Today you cannot get anywhere near the stones with large fences all around the site and access to the site restricted to tour groups only and within certain areas only. Not far away from Stonehenge was Amesbury, recently confirmed as the longest continuous settlement in the United Kingdom. Amesbury, including Stonehenge, has been continually occupied since 8800BC; following an archaeological dig which also unearthed evidence of frogs' legs being eaten in Britain 8,000 years before France.

Amesbury's place in history has also now been recognised by the Guinness Book of Records and blows the lid off the Neolithic Revolution in a number of ways, providing evidence for people staying put, clearing land, building, and presumably worshipping monuments, and in many ways was a forerunner for what later went on at Stonehenge itself. The first monuments at Stonehenge were built by these people. The River Avon, which flows through the area, was a highway for travel between settlements.

Again close to RMCS, just 30 minutes' drive, was Avebury with its 4000 year old ring of Neolithic stones where once again we could walk round and touch them. I am unsure if that is possible today but doubt it.

Another drive took us to the White Horse a carving into the limestone on the escarpment of Salisbury Plain, approximately 2.5 km east of Westbury in Wiltshire and an hour's drive from RMCS. Lying just below an Iron Age hill fort, it is the oldest of several white horses carved in Wiltshire and is over 400 years old.

The horse is 55m tall and 52m wide and has been adopted as a symbol for the town of Westbury, appearing on welcome signs and the logo of its tourist information centre. It is also considered a symbol for Wiltshire as a whole.

Keiran and Carol soon joined in the college life and both started to take riding lessons. The college had horses they could use and some of the cavalry officers on staff had brought their own horses and grooms with them and it was these grooms who were the riding instructors. Not a bad posting for a young trooper, two years of pampering horses and nubile young girls.

Over Easter we decided to hire a canal boat and do the Oxford canal system down to the Thames and back. A typical narrow boat scene is shown

There was a canal boat company operating out of Abingdon not far from us and we hired a narrow boat for the Easter weekend. This type of holiday in England especially is a very relaxing and interesting way to see rural England at a leisurely pace and with no time pressures other than those you impose upon yourself. Our boat was 55 ft long and there was a double cabin, a 2 berth cabin and a bunk forward that filled our needs.

The narrow boat nomenclature is very accurate as the beam was about 7-8 ft. It had a traditional stern i.e. room for helmsman only with seating in the bow to take in the passing parade a saloon in the stern with galley including gas stove and fridge providing seating for us if we ate aboard. Usually, however, most meals were taken at the many canal side pubs that we encountered as we drifted along at walking pace through the canals. In fact it was not unusual for Carol and the kids to walk along the tow path as I drove the boat.

I had been very particular about the kids sitting down whilst we were travelling and waiting to be handed on to the path from the boat when we stopped. David being David thought Dad was a dill and at one stop proceeded to jump off the boat as soon as we got close. Naturally he misjudged the leap and was heading for a fall between the iron sided boat and the canal side. Dad, the dill, was watching and preparing to hand him over to Carol who was on the towpath waiting so it was simply a case of grabbing the collar of his Paddington coat and picking him straight up, his feet never even touched the water.

Most locks were manned and we only had a couple we had to do ourselves. There was one obstacle that did cause concern though. There are a number of walking paths that meander through the countryside and when they reach a canal they are crossed by a bridge that is, obviously, a barrier to canal boat progress. These bridges are counterweighted and it is simply a matter of stopping the boat short of the bridge pulling on the rope and lifting the bridge up vertically and driving through and lowering the bridge after you are through. This worked except for one memorable occasion.

Carol would never drive the boat so she had to lift the bridge. I stopped short and she went and did the lifting bit. But one time as I was driving the boat through, gravity and the reality of disproportionate weight on a lever saw Carol being pulled into the air by the bridge rather than her weight holding it open. I saw the problem when we were half way through and went full steam ahead which did not result in any whiplashing acceleration just a steady increase in speed. I ducked as the bridge fell and it missed the cabin top and me but struck the push pit rail on the stern but did little damage. "Very lucky."

Carol was white with shock and I could never get her to open a bridge again, I used a length of line on the boat to tie it off when it got vertical, drove through returned lowered and untied the rope tedious but a lot safer.

On Easter Saturday we stopped overnight in front of a pub, as usual and we noticed there was to be a competition between two rival pubs the next day, Easter Sunday. We decided to stop and watch the frivolity. Each pub turned up with their own band, one on each bank of the canal and soon there was music echoing out across the canal sometimes playing the same tune sometimes, not. A lift bridge just downstream of us allowed you to walk from one side to the other. We being moored up between the bridge and the pub had a ringside seat. Not surprisingly the pub was doing a roaring trade and now it was time for the main event. The worthies of 'our' pub had been challenged to a tug-of-war with 'our' side on the pub side of the canal and the enemy on the other. A rope was passed over and the trick was to pull the opposition towards and, hopefully into the canal. Despite the clamour by the boys to get involved we stayed as spectators and just as well for as should be expected soon everyone was in the canal, 'Winners' and 'Losers'.

During the afternoon we waved farewell and continued on a few hours to a new pub location and had a relaxing time away from the rowdy mob just down the canal. Easter Monday was spent completing the trip and handing back the boat. A fun weekend!

******

Caro 1 was to discover a problem with the MQ around about now, "Harry," she cried, "the cool box is not working. The butter is melting." The fact that the ambient temperature outside our MQ had risen had escaped her notice and she can still be made to blush when reminded.

Chapter Thirteen - Touring Europe

During the mid-year break between semesters Carol and I decided to take a tour of Europe. We found two wonderful places to accommodate the children while we were away – Keiran went to a riding camp outside Taunton in Somerset and the two boys to a Nanny School at Hungerford not far from where we lived. Both turned out to be very good choices and I was not sure we were going to get them to come home when we went to pick them up. The boys shared a bedroom but had a trainee nanny each so were suitably spoiled. Keiran was involved in all aspects of the 'horsey' life, riding being the least of it, with animal care and grooming and tack maintenance an important part of the experience. She loved it.

I had remained in contact with the Belgian Army Commandants, Resmet and Le Beau in Leopoldsburg and Tony Hughes-D'aeth in Munich so arranged to tour around starting and finishing at Ostend and going to Leopoldsburg/ Brussels, Nancy in France, en route to Lake Como, then Kitzbuhel near Niederau in Austria, where we stayed for a week, before driving back via Munich.

The local travel agency looked after all the bookings and all we had to do was pay for it and pack the car. We had decided to drive our car rather than hiring one in Europe and economically it was sensible but driving a right hand drive car on Belgian, French, Italian and Austrian side roads was a challenge at times. Fortunately the majority of our driving was on autobahns or equivalent so we survived.

After taking the children to their respective accommodation we drove to Dover, about 200 km, and caught the overnight ferry to Ostend and drove to Brussels, 120 km. I had arranged accommodation at Club Prince Albert, the Belgian Army Officers' Club, where I had spent a night the year before. It is close to the Grand Place and an easy place to park. The next morning we drove to Leopoldsburg and met up with my old friends from the previous year. We had a lovely day with Madame De Smet and her son taking us to a park where the Belgian government had established a 17ᵗʰ century village by collecting the old cottages, barn, shops and blacksmiths shops from around the country and bringing them to one place. It was a fascinating place and we were able to see ladies making the lace work that Belgium is famous for on old 17 century looms. After a pleasant dinner we drove back to the club and prepared for a day of sightseeing in Brussels. It was a lovely day and we had a good time simply walking the familiar, to me, sights of this pleasant city. We did, of course, rub the nose of the dolphins near Manikin Pis but as yet it has not worked its magic and we have not returned. One of the sights that I did ensure Carol saw was the ladies in the windows that we used to pass each night when returning to our hotel the year before. This was followed by a small drink at our favourite bar 'Harry's Bar'.

The next day we left on the 375km drive to Nancy. There was no reason to choose Nancy except that it was on the way and we simply had dinner, a sleep and breakfast before taking off for Lake Como, 500 kms away over the Alps. We went via the Dolomites and passed briefly through Switzerland. It was a warm day and I was enjoying the fitted air-conditioning when it failed. The remainder of the trip even though we were climbing was unpleasant but I wanted to reach somewhere where I could find linguistic assistance when I sought a mechanic's help.

We arrived at our pensione in Lake Como in late afternoon and were so dry that we simply dropped our bags in our room and went to sit on a balcony and consume a number of much needed drinks, Cinzano and soda for Carol and Lowenbrau for me.

The friendly staff prepared a lovely meal for us and served it to us while we enjoyed the cool breezes and cool drinks. I had a sardine salad and the fish were cooked whole. The waiter used a fork and spoon to fillet the fish and serve it at the table. We were most impressed. Morning saw me concerned about the car's air conditioning and I had a quick look before seeking outside help. It was lucky I did, because the problem was very simple and easily remedied. Being clumsy of foot I had kicked the cover on the fuse box just under the dash and dislodged one of the fuses, the air-conditioner fuse. All I had to do was replace it in its holder and we were right to go. Stupid!!

With this major repair completed we loaded up and, with air working, drove the 350 km through Niederau and 30 kilometres later arrived at Kitzbuhel a small picture postcard of a village. In June the lower levels of the Alps are beautiful and popular with tourists interested in walking and taking in the spectacular scenery. We had a self-contained chalet with a large deck that looked out on the now grass- covered ski slopes. Down valley about a km or so was a ski lift which took us up on the ridge opposite our chalet and from the top of this lift we could walk along a flat road with a view into the next valley, to another lift a few kilometres away and then take that lift back to the valley floor and along to our chalet. There was an inn just a short walk away and one night we went there for dinner and met up with an Austrian couple. Once again it was a case of linguistic ability improving with the amount of liquor consumed and we ended the night very late after dancing and drinking and cementing Australian-Austrian relationships.

We stayed for a week at Kitzbuhel and apart from a day trip to Innsbruck were content to live the village life; walking into the picturesque little village each day and relaxing and watching the hang gliders swooping down from the ridge opposite to land in a field in front of our chalet. I was tempted but common-sense prevailed and I did not have a go. The weather during our stay was perfect and it remains one of our best holidays.

Our final day arrived and we packed up and drove the 250 kms to Munich and stayed with the Hughes D'aeth's for a couple of nights. Sue and Tony arranged a dinner with Dirk Hanson and his wife who were to go to Australia as the Krauss-Maffei rep the following year. Tony noted that I had Kangaroo decals instead of the traditional three letter country codes that adorn European cars so he had to have some. I told him I had a spare set and sent them to him when we got back to Shrivenham.

Sue drove us into Munich the next day and we arrived in Marienplatz in time to watch the Glockenspiel do its thing Sue left us to roam around and we visited the Hofbrau House and Carol asked for a small glass of beer that did not translate well and I had to finish off her stein and mine. "The things we do for love."

We took a taxi back to their house and had a final night there before leaving the next day to drive the 800+ km to Ostend. Time was on our side so we overnighted at Club Prince Albert in Brussels on the way. That allowed us a leisurely breakfast and quiet drive to Ostend. I had been taken to a fine seafood restaurant outside Brussels by Maj Brunin, so we stopped there for lunch and I revisited the exquisite Coquille St Jacques that had been my choice a year before.

Arriving at Ostend with plenty of time to board the ferry we walked along the shoreline and found vendors selling pickled mussels from small stalls. I love mussels so had to have a serve.

The trip back to UK was uneventful; the best type of sea trip as far as Carol was concerned and after we cleared Customs with our duty-free goodies we drove to Shrivenham getting back early in the Saturday afternoon. We drove to

Taunton and Hungerford the next day and retrieved our children. They did not appear to have missed us much, having had the time of their lives.

****

Our second semester began with a rush of visits and papers to be written and it was obvious that the next few months were going to be full on. This was the time where we had to knuckle down and work on our project and presentation but it was also the start of the short rugby season. RMCS 'Owls' competed in a regional competition against police, fire brigade and our sister college UK Staff College Camberley, plus a few others. I think we played them all only once and most games were held at RMCS.

I made the team despite being mocked by some others with, "Fat Harry is going to play rugby." now Fat Harry had put on a lot of weight but I was sure a fitness program would right that so started to run the blocks and fields losing a little at a time. By the time the trials were to take place I was probably not-quite-as Fat Harry and when the coach lined everyone up for a sprint up a slope turn around the tree and return, the sniggers had died away.

I played centre and had a reasonable season scoring the odd try but on the last game I tried to tackle my opposing centre, a good player from Camberley, went down clumsily and broke a bone in my right ankle. For some reason we had no reserves and I went to the left wing and tried to hide but the ball kept chasing me and twice I had to kick it out using my only kicking foot the injured right one. "Mothers tell your children to learn to kick with both feet."

Shortly after the game a visit to the RAF Hospital saw me with a cast on my right foot and an inability to get around easily.

160

Unfortunately the next visit we made was to the Armoured School at Boddington and as part of our package we were to experience live firing, with three rounds from the main gun of the MBT Chieftain and a number of rounds through other armoured fighting vehicles (AFVs). I had to pass them off to others as I could not climb into the tank or the AFVs. These very expensive and thus rare experiences are called 'Yippee shoots' and are eagerly sought. "STIFF!"

National Days. Naturally each foreign country celebrated its national Day and the Canadians invited the other Canausters and some Brits to a street party on Friday 1 July, Canada Day, where the drink was Moose milk.

Moose Milk is a traditional Canadian alcoholic mixed drink with roots in the historic celebratory events of the Canadian Armed Forces. The Canadian Navy, Royal Canadian Air Force, and Canadian Army all claim to be the originator of the drink.

Moose milk is composed of five different classes of ingredients:

Hard liquor: typically a combination of Canadian whisky, vodka, or dark rum

Coffee beverage: Kahlua and occasionally prepared coffee

Dairy: a combination of whole milk, cream, condensed milk, egg nog, or ice cream

Sweetener: maple syrup or sugar

Spice: nutmeg and occasionally cinnamon

Egg yolks are sometimes used directly or indirectly through egg nog or ice cream to prevent separation of the drink.

It was drunk by the ladle full out of a bucket. Maple covered wings and other things were barbecued and as we only had to walk home it was a fun day.

However, as we had missed 26 January the Australians had a day in October set aside for celebration and it was to ensure this day was memorable that all Australians had contributed each month to the party fund.

I was not on the Australia Day committee so have no idea how much was raised but considering we laid on a lunch for nearly 300 people it must have been a fair amount. Our guests were all of the Technical Staff Course students and all of our instructors and their partners. the committee had organised a lunch that was catered by the RMCS staff and include BBQ angus beef carcasses brought down from Scotland and butchered on site, a seafood extravaganza with garbage tins full of prawns simply set out around the party area inviting all guests to help themselves, a salad bar and specially imported fruit. Just in case someone wanted a drink to wash it down there were cases of beer and Australian wines available, plus the obligatory fruit juices and soft drinks.

I well remember talking to the wife of one of the senior staff, a full colonel, and she admitted that apart from mince this was the first meat they had eaten since returning from their Army on the Rhine (AOR) posting in Germany. Thankfully a year or so after we left all UK services got a massive pay rise and that ridiculous situation was changed and people who followed us did not have to hide their relative wealth.

Chapter Fourteen – Return to Australia

Our year was coming to an end and everyone was concentrating on our next postings, the majority, us included, were going to Materiel Branch AHQ Canberra, although two went on to further training in the UK. We were all busy packing up, arranging removals and arranging our travel. You were given plenty of flexibility in determining how you would get to Canberra. The criteria were that you had to be there by a certain time and if you stopped on the way it was to be taken as leave and if you exceeded the nominal stopover period (2 nights) you had to pay the extra hotel costs. Basically your travel allowance covered a direct return with a two day stopover, anything over that was your problem. We decide to return via Hong Kong and stay an extra couple of nights. We were booked into the Holiday Inn in Kowloon and had a suite of rooms. By coincidence another Australian couple had done the same thing and we later had Christmas day with them.

The first day after we arrived we decided to go for a walk through a market, a mixture of outside stalls and small shops, opposite and with kids in tow went to experience the mysterious orient.

It was an interesting medley of colours and aromas and we were delighted to be a part of this varied and wonderful part of the world, that is, until we looked around and did a head count and noted we were one down. Jamie was missing and we launched into a frantic search but despite the fact that a lot of the stall holders and shopkeepers could speak English we could not find him so decided to return to the hotel to get help. When we arrived the first thing we saw was a smiling Jamie. A man had noticed this blonde headed 7 yo roaming around looking lost, and luckily Jamie could remember the name of our hotel and he was taken back and left with the receptionist. I wish we could have found the Good Samaritan but he had just left without even leaving his name.

Jamie was probably trying to catch up in the 'Hey, look at me stakes' with David who by this time had met Mickey Mouse, had his own nanny, been run down by a BMW and nearly fell into a canal, not a bad year for someone so young. If this was so he chose a heart stopping way of doing it.

The meals at the hotel were good but expensive so we sought an alternative. One of the Chinese stalls opposite the Holiday Inn served breakfast and the kids thought it was special to eat a Chinese breakfast each day. After the inner kid was satisfied it was time for retail therapy for the adults and to do that we made our way to the large shopping centre across on the island.

We only had to catch a Star ferry and walk a short distance and we were welcomed in for a nice and cheap way to spend a couple of days' time and a few travellers' cheques. We ended up with tailored shirts for me and boots, blouses and handbags for Carol. The kids received a fine collection of Christmas presents that we tried to hide until Christmas day but they found them in our room, and we had to bring Christmas forward.

The Star ferries were worth the effort on their own for even though it was only a short trip we could see a lot of activity on the harbour as we crossed, and on board it was a snap shot of the real Hong Kong. For a small fee you could sit on the upper deck in comfort and for half the small fee stand on the lower deck with the rest of the crowd.

The hotel had a babysitting service which enabled the Warks and us to have a seafood extravaganza dinner on the Jumbo Sampan restaurant moored in the harbour.

After Christmas we left on our final leg back to Australia flying Singapore airlines to Perth. Win, Aunt Stella and my half-sister Joy met us at the airport and we spent some time with them before flying on to Canberra, arriving about 28 December.

I did not have to report in until 4 January and with our removal from UK not expected for a few months we looked forward to an extended stay in the Park Royal Hotel/Motel in Northbourne Avenue, Canberra.

There were plenty of minor details to keep us occupied though firstly where were we going to live and flowing from that where would the kids go to school. Having sold our car before going to UK it was time to think of getting some wheels as well. Obviously we could expect to be high on any list of those waiting for a MQ but we decided that it might be time to buy our first house. We knew nothing about Canberra so had no preconceived views.

Chapter Fifteen - The First House.

New Year's Day was on a Sunday that year and we decided to start looking in the real estate advertisements on the Monday Canberra Times and tried to get a feel for what was available and where. It was an interesting morning comparing the properties with a Canberra road map to make sure we were not looking at something miles away from work at Russell Offices. I basically drew a circle about 16 km radius and saw what was available and how much you could expect to pay. The basic requirement for us was a 3BR fenced property with a reasonable sized yard and off street garage parking. Apart from that it needed to be reasonably close to a primary school and a decent shopping centre. In those days Canberra was a small city with three main shopping centres, City Central, Belconnen and Woden, with Belconnen and Woden equally distant from Russell and within the 16 km radius. City central was only a few kilometres away and there were many nearby leafy suburbs from which to choose. The reality of these suburbs, of course, was that they were also among the most expensive.

One real estate agent had a number of potentially suitable properties and I decided to give him a call and try to arrange a look at some he had on his books. It was a public holiday but that did not slow him down and he agreed to pick us up from the Park Royal where we were staying waiting for our goods and chattels to catch up, and drive us around showing a few of his listings from the street to help us orient ourselves and perhaps spot something we might wish to inspection later.

The tour took us in a big circle showing us the better suburbs close to the city and then out to Belconnen before swinging back to Woden. The older, better suburbs were fine but too expensive so we resolved to look closer to the outlying shopping complexes of Belconnen or Woden. Of the two we

preferred Woden so made arrangement to have a look at a
few houses on that side of Canberra over the next few days.
One area that he wanted to show us was the soon to be
developed suburb of Tuggeranong. At that time it was no
more than a wide open space with a few crossing streets and
the promise of much development in the years ahead.

Pioneering was not our forte, we wanted established.
So, although the pricing was right, there were few houses to

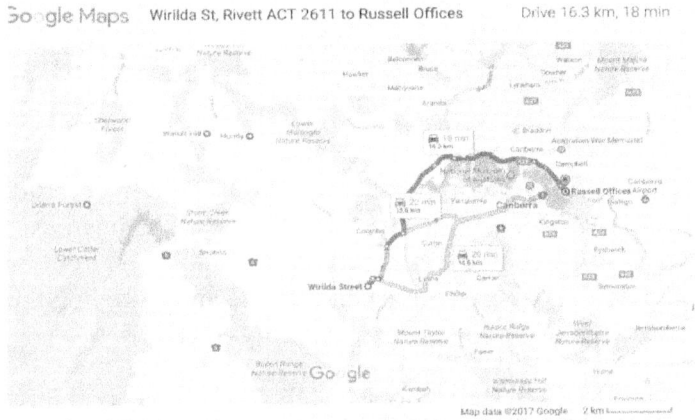

view and thee was not much in the way of infrastructure.

Eventually we found a three bedroom brick and tile
'ex-govie' in Rivett, 25 Wirilda Street, about three or four
kilometres outside Woden making it right on our 16 km

radius.Rivett to Russell offices, See above

This one had been extended and had a brick paved courtyard out the front and an established lawn in front and around the side of the house. It was newly painted in white and had a tree in the front courtyard and another out the back behind the garage giving some shade and softening the brick courtyard and walls.

It was heated by oil, which was all the go in the late 70's and cheaper and easier than wood and much cheaper than electricity. We paid $35000 for it which was close to the median price of houses in Canberra at the time $37000.

As luck would have it I had saved sufficient for the deposit by banking our UK allowances and living off our salary whilst at RMCS. This was to have a second benefit as the real estate agent brought to our attention, while we were signing the papers, that while we had been away there had been a government incentive payment of $2400 for first home buyers who had shown they could save money by depositing regularly over a period of time, 18 months. Our regular deposits, therefore, qualified us for a percentage of that - $1600. This largesse paid for the stamp duty on our house purchase which was most welcome. With our defence service loan of $25000 we needed a second mortgage of about $7000 and Carol decided to see if she could find work given that all the kids would be at school.

Now that we had a home to base our activities around we were able to look at schools for the kids and it was easiest for them to go to the local Rivett primary which was just a short walk to the end of the street and across a park and they were there. The house settlement was delayed for 2 months and this meant we had to continue our time in the Park Royal

and when the school year started it was a case of taking them to school and leaving them in day care and picking them up from more day care after school.

Chapter Sixteen – Mat Branch - Year one

My posting to Materiel Branch began on Wednesday 4 January 1978, and I fronted up with my orders, but as no one was around I was told to report in on the following Monday 9 January. My first day was spent being photographed and processed by the Camp Commandant and issued with my security card. This card needed to be carried at all times.

I found my way eventually to my office and learnt I would be sharing it with Col Dobie whose house we occupied in RMCS on their return to Australia. It was a pleasant surprise and we and our families became very close friends.

Desk officers in Materiel Branch are responsible for the staffing of the Army's request for new equipment. To put it very simply, it all begins with a ministerial white paper on Defence that outlines the threat and from that the services determines the force needed to meet that threat. From this paper flow bids for manpower, equipment, training and funding. In Army, the Operations Branch considers the bids and with the various arms directorates, infantry, artillery, armour etc. recommend a particular weapon system to the aim of the white paper.

All aspects, the weapon system equipment, the personnel to man it, the ability to maintain it in the field and the training to enable it to be operated and maintained, have to be considered and have equal weight in the decision process

The Materiel Branch part of this is to address the equipment question by taking on board requests for a particular piece of equipment determining if it exists in, or is being offered to other armies, and if so writing a paper comparing the already existing or proposed item against the weapon system parameters requested. These weapon system parameters requested by an arms directorate and endorsed by Ops branch were referred to, by us, as wish lists and not concrete criteria. It was our task to meet the parameters as close as possible with in fiscal limits. If we had allowed ourselves to be locked in to the Ops Branch parameters as concrete criteria then we could have been looking with blinkered eyes at something a staff officer had noticed in Janes and not the range of possible weapon systems that would meet, and perhaps exceed the parameters.

Within Mat Branch, Staff officers had responsibility for a class of weapon system or ,if a major piece of equipment, that item only. I was posted into the electronic weapon systems section and had to cover the Ground surveillance equipment. This included anything that could detect the approach of an enemy to our location, Radar, sound sensors, ground sensors, thermal imaging, infrared and night observation equipment, and drones. The primary user of these weapon systems were Special Forces so I spent a fair amount of time in discussion with Col Mike Jefferies who had been promoted to colonel as the first Director of the Army's Special Action Forces in 1979.He was instrumental in developing the surveillance concept for Northern Australia and as Director of Special Action Forces he prepared the development of the Australian counter-terrorist concept and capability. He was later promoted to major general and then became governor general.

Col who shared a room with me was working on the introduction of an intelligence processing computer. To assist us we each had a public servant technical officer who had valuable corporate knowledge of the machinations of the staffing process as well as sound technical knowledge on the weapon system types. The vehicle used to carry the process was called a Major Equipment Submission (MES)

****

An MES was a paper that embodied the operational criteria and the selection process to arrive at a preferred solution and then the fun begins with the need to develop a plan to procure the selected weapon system. In the 1970s there were no Excel spreadsheets so all the elements of the weapon system the prime cost, the currency rate, the number of operational equipment needed, the reserve and training equipment, the repair parts, test equipment, publications, and stores, were detailed and costed to determine the overall bottom line cost that had to be passed to the Minister and included in the budget. This was a far more difficult task then than it would be now. Any change to the base numbers, prime cost, currency rate and required equipment numbers, caused changes in every element of the spread sheet due to the fact that each of the ancillary elements were based on a percentage of the prime cost, e.g. repair parts might be 40 per cent of prime cost and so on. These changes had to be recalculated manually and inserted, into the spread sheet then the whole lot added up both vertically and horizontally to arrive at the new bottom line.

The MES could cover a 3-5 year period from start to final procurement and was therefore a significant document. Each year it had to be massaged and amended to ensure an accurate and fully staffed Submission was ready for the minister to include in the budget projections. The term staffing needs some explanation; once the MES is ready it is distributed to a number of interested parties who read it and comment on it from their particular point of view. About 20 plus copies were produced and this in itself was a logistic exercise, and given that a copy could have no error, typo or whatever, we spent a lot of our time moving between our offices and the typing pool to arrive at a perfect copy that became the principal and all reproductions were made from this copy.

Each MES could run to at least twenty pages some much more including annexures and spread sheets and they had to be hand collated and stapled together ready for distribution. The method of collating and distributing was almost Dickensian. To use our modest submission as an example, if there were twenty pages in the MES and a distribution list of twenty it meant we needed twenty piles, one for each page, and there would be twenty or more in each individual pile. They were placed in order around a long table and all our sections and usually a couple from the typing pool would start at one end and walk around, one after the other, picking up a sheet from each pile until you reached the end where someone would staple the now complete MES, tick the next name on the distribution list and place it in a box ready to be taken to the distribution centre who sent them out. This was fine for a normal staffing distribution but each year we had to do the ministerial distribution and, for this one; forty-seven copies had to be copied, collated and distributed. Obviously for the much larger submissions the task rivalled that of the Aegean stables.

Although, the MES was our primary focus but we also had to keep ourselves up to date on advances in our areas of concern and that meant we spent a good proportion of our time schmoozing 'gun –runners' as we termed the equipment suppliers' representatives who regularly contacted us to press the claims for their company's latest and greatest bit of kit. Depending on the sources of a weapon system that fitted your MES you might have a short visit to your office and a cup of coffee that you provided This was the case if the source was local and governmental, e.g. Engineering Design Establishment (EDE) in Melbourne or The Weapons research Establishment (WRE) in Elizabeth outside Adelaide. If, however, it was an international company then you could expect to be taken for lunch, or dinner, or both.

Col explained the subtle difference of accepting a bribe and accepting entertainment and recommended that I did

173

[Type text]

what he and a number of others did and join the Press club and sign the gun-runners in as your guests after that the payment of the account was between you and the 'gun-runner.' No one, that I am aware of anyway, was silly enough to take money or gifts, other than the occasional plaque that you hung in your office. I think in the first year because of the type of weapon system I was responsible for I was taken to lunch about twice by visiting 'gun-runners'', the same one each time.. Most of my entertainment 'perks' were when I visited places like WRE for progress meetings or demonstrations and I was on travel allowance. Hardly a private helicopter flight or visit to the polo.

Col and Marlene and their two boys lived in Chapman about ten km away and a more recent development with nice red brick and tile house set in nice gardens. Our families became quite close and we spent many a day or evening together.

By applying through an employment agency Carol got a job with the public service in consumer affairs and we settled in to a comfortable life style.

Sport in Act. The park at the end of Wirilda Street had a couple of Rugby grounds and our boys were introduced to the sport through the neighbours, both opposite and next door, whose boys played. Many a football match was played on ours and the next door's front lawn so it was a simple thing to enrol them in the Warriors Under7s and Under 9s. Warriors were one of a number of junior teams that operated under the guidance of Canberra Royals Rugby Union Club. I joined Royals as a social member and when they were looking for coaches took over the two teams the boys were playing in. In the U7s, the ground is smaller and the coach is on the field directing traffic –good fun. On one occasion, though, I had to stop the game and ask a parent to leave as he was ridiculously abusive of the opposition players and even worse his own son. He muttered his way off and we continued. His boy kept playing out the season and the father did return to watch but did not apologise. More importantly, however, he kept his offensive opinions to himself.

Both our boys were very good players with plenty of pace and determination. Jamie was asked to go to Sydney for what was an annual series of games against a team from Warringah. The boys were billeted out and the parents stayed in a motel. It was a special weekend for both of us.

Rugby in the ACT at that time really meant rugby union as Rugby League was a much smaller sport and apart from the Queanbeyan Rugby League Club, club-house, an edifice built on poker machines there was not much interest among our circle. We did, however, use the club regularly and play the pokies. That was to change some three years later.

The 1978 Wales rugby union tour of Australia was a series of nine matches played by the **Wales national rugby union team** in Australia in May and June 1978. The Welsh team won five matches and lost four, including losing both of their international matches against the **Australia national rugby union team**. On 13 June 1978 Wales, who were **Five Nations** champions at the time, played ACT as part of this tour. ACT had trailed at half time, 6 to 16, but came back and won with a penalty kick by 18yo Michael O'Connor (see below) the Royals centre in the final moments of the match. We all went wild and it was a late night at The Rugby Club in Barton that night.

One of those who played was reported as saying" "When we won, the crowd came out on to the field and ran to us; they wanted our autographs and not the Welsh. We couldn't believe it. The Welsh had some big names and we couldn't figure out why they wanted us. The Welsh players didn't want a bar of us, they had the shits, especially with someone like the ACT beating them … it's a great memory."

175

Michael David O'Connor (born 1 February 1960 in Nowra, New South Wales) represented Australia in both rugby Union and league, in other words a dual international. Originally from Canberra he toured with the undefeated Australian Rugby Union Schoolboys tour of Great Britain and France in 1977 alongside Ella brothers Mark, Glen and Gary, and Queenslander Wally Lewis. His club rugby career was with the Canberra Royals He later played for the Wallabies in 13 Tests from 1979 to 1982 and then the Kangaroos (Australian Rugby league representative side) in 17 Tests from 1985 to 1990.

****

Having bought a new set of golf clubs in UK I joined the Queanbeyan Golf Club and played regularly there on weekends and played with the Army Golf club on Wednesdays visiting all the other ACT clubs in rotation. My golf did not vary over that time and I stayed about a 16 handicap very rarely breaking it but being pretty consistent.

One of the junior pennant players at Queanbeyan, a 16 yo who was a fair rugby league footballer, gave up rugby to play golf. In 1978 he won the ACT-Monaro Schoolboys golf title and we determined to watch his progress with interest. His name was David Campese and he was to make a name as one of the greats of rugby union. David Campese played his first game of rugby union for the Queanbeyan Whites in 1979 in fourth grade. During 1980 he was promoted to first grade. After two years of first-grade rugby, in 1981 Campese was promoted to the Australian under-21 squad to tour New Zealand that was beaten 37-7. He ended up playing over 100 tests for Australia and scoring 100 tries.

The climate in Canberra finally got to me, too damn hot, and I decided to resign my membership of Queanbeyan in 1979, and only played army golf from then on, on Wednesdays.

****

Col introduced me to a group within Mat Branch that were in his year at RMCS and they had set up a poker school that rotated between houses once every three weeks so I joined in and this became a regular member during our time in Canberra. Two of them Brian McCauley and John Sheedy were to have a significant influence on my life in a few years' time. Someone had a brainstorm that our poker school should lease a 2 yo filly and race it locally. The filly was called Likely Lass and on her first outing on sand at Goulburn she won well. Another run on grass at Canberra saw her finish fifth but we were happy enough to put her out ready to bring her back as a 3yo. Unfortunately, the dill of a trainer had her doing fast work on an icy track and she fell coming round the bend into the straight and although not seriously hurt she would never take a bend with confidence again and after a year's trying we bailed out.

Another thing that I found when I got to meet up with the rest of the Mat Branch officers was the fact that there were two little cliques; one that did the Canberra Times cryptic crossword and the other that played mah-jong at lunch time.

One of the driving forces in this clique was an Education corps officer, John May, who I had served with at Army Apprentice School at Balcombe, some 13years before. The first thing on his agenda each morning was to buy the Times and copy the crossword a number of times and distribute them among the faithful.

We would work away and take a mental exercise break every now and again by trying to work out a clue. By morning tea time you could expect to have some done and it was usual to ring around and see if anyone else had more than you. Thus it was more a collegiate effort than individual but it was very good brain food even if solving some of the clues made you feel you had to turn your brain inside out to see the other side of reality.

John was also the custodian of the mah-jong set so we used to gather each lunchtime, in his office, to play. It was a fast and furious version with no scores recorded and the aim simply being to go mah-jong. I stopped playing after leaving Canberra and when I tried to play a few years ago I had to start again.

Towards the end of the year Col introduced me to another Russell Office's group. The Russell Offices Hash House Harriers and I was soon joining in every Monday evening running around the streets of Canberra yelling "on on" with the best of them as we 'hounds' once again picked up the chalk mark lures laid by the 'hares' earlier that afternoon.

The run usually went for 40 minutes and finished at a park near where we had parked our cars and started off. There was a 'Hash Bucket' waiting which was a container filled with a mix of beer and ginger beer and we would dipper out mugs full as a reward for completing the run. I often wondered what suburban Canberrans thought of the sight of thirty to forty, 30-40 year olds running past blowing horns and shouting "on on". I made a number of friends from that group that were to be a significant influence on my later life when I caught up with them again on my next AHQ posting in 1981.

Rivett Primary School had an annual 'bush week', in Monday out Friday, where a couple of classes, 5 and 6 grade, were taken to a bush retreat where there was a large covered deck and kitchen setup, big wood stoves and fridges, some toilet blocks and a couple of cabins adjoining a large grassy field; the kids had tents and sleeping bags and the teachers the cabins. I volunteered to help and arranged some army

[Type text]

tables and chairs and tarpaulins to provide more shelter and comfort.

I also took on the job of camp helper and helped the kids with their tents and setting up wash points near the toilets and cooked piles of toast and made sure there was cereals, milk and fruit set out for the kids on a help yourself basis controlled by the teachers, then a breakfast for the teachers. After breakfast the teachers took groups out in the bush and by playing tapes of bird calls tried to entice as many birds as possible into that area and the kids had to draw them, write a description and when back at the decked area search through the books to identify the birds. I was fascinated by it all and occasionally went out and sat with a group and watched and listened with them.

About 1600hrs I would start making dinner and cooked for the multitude usually a BBQ or one pot meals with a few side vegetables and it seemed to go down alright. Dessert was easy tinned fruit and ice cream. After the kids were out of sight I allowed myself a couple of beers with whichever teacher was not supervisor that night.

****

The midyear school holidays in the Act came about the end of June so we started a family tradition that was to go on for years. Reading the Army newspaper one day, I saw an ad for holiday accommodation for soldiers and families in Army owned resorts at various locations around Australia including Terrigal in New South Wales, Rottnest in Western Australia, Gawler in South Australia, South Arm in Tasmania, Queenscliff in Victoria and more relevant to us Bilinga in Queensland. Bilinga is a seaside suburb on the Gold Coast, mid-way between Kirra and Tugun and the Army Holiday resort was next door to the Coolangatta Airport and only a couple of hundred metres from the water's edge.

It became our habit to pack up the Escort and drive up there for a week during each school holiday break in June and/or September. We would leave Rivett about 0400 hrs and stop for breakfast at the start of the New England Highway at 0800 hrs. Carol used to make sandwiches the day before and freeze them overnight so they were ready to eat fresh the next day; our first day of travel.

Depending on how we felt on leaving the breakfast stop it was about 4.5 hours' drive (410 km) to Kempsey or about a 6.5 hours' drive to Glen Innes (600 km). I sat on about 100kph with the aim to do 90 km every hour including stops. The first choice left us with another 4.5 hours' drive on day two or we could make a shorter 4 hours' drive (370 km) it did not matter much as we usually had a breakfast before we left on day two and arrived at the accommodation mid-afternoon.

This gave us time to stock the fridge with beer and the shelves with groceries and take the kids for a swim in the resort's pool. We all loved the break and one extra special treat was a McDonalds as at that time there were none in the ACT. Another treat, however, was for the kids to select one of the theme parks and we would make a day of it and it was up to them which one it was. I must admit with only 'Movie World' or 'SeaWorld' and later on 'Wet and Wild', to choose from it was not a big decision.

Chapter Seventeen – Mat Branch Year 2

At the end of 1978 I was given responsibility for a different MES. This time the prime equipment was a mortar locating radar to replace the current aging AN/KPQ-1 mortar locating radar that had seen service throughout the Vietnam War.

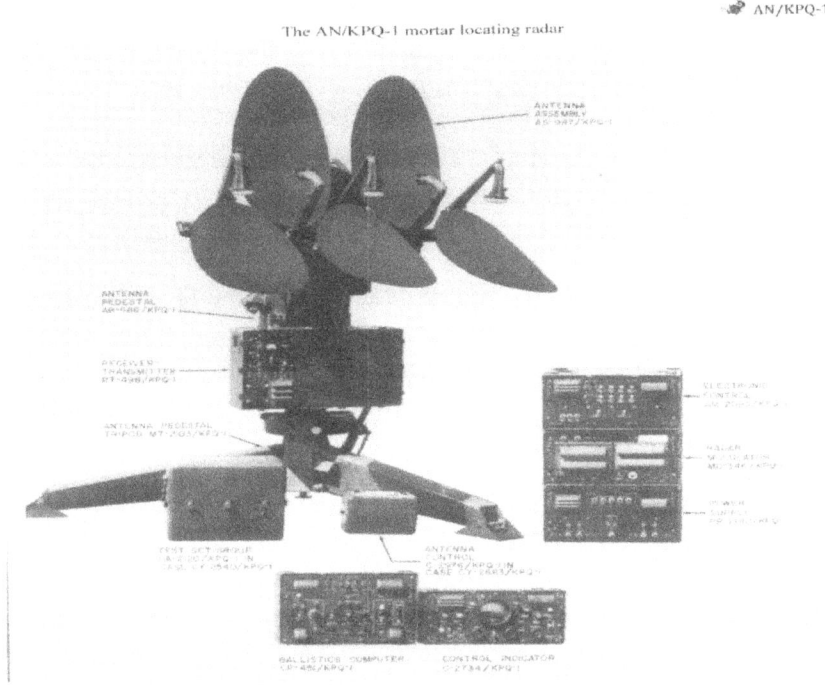

The AN/KPQ-1 mortar locating radar

AN/KPQ-1

Mortars, using indirect fire, became a major threat to infantry in World War II. It was, however, found that mortar bombs in flight could be detected and tracked by radar. USA and UK anti-aircraft radars were used and some specialised mortars locating radars appeared at the end of the war, and were used in Korea with varying degrees of success. Hostile mortars had to be accurately located before they could be attacked with indirect fire from guns or mortars. Since hostile mortars moved frequently to avoid return fire it was essential

181

[Type text]

to have a means of locating them to a few tens of metres of accuracy and to be able to respond quickly when they are located.

The aim of my MES was to find and process to procurement and introduction into service a counter-battery radar. A much more advanced capability than that required of AN/KPQ1.

Counter-battery radars or Weapon Locating Systems are mobile **radar** systems that detect **artillery** projectiles fired by one or more guns, **howitzers**, **mortars** or rocket launchers and, from their trajectories, locate the position on the ground of the weapon that fired it. More advanced systems can electronically send aiming instructions to friendly artillery for firing at hostile targets with **counter-battery fire**.

Modern counter-battery radar can locate hostile batteries up to about 50 km away depending on the radar's capabilities and the terrain and weather. Counter-battery radar is attached to an **artillery battery** or their support group.

This leap in capability was made possible by the invention of the Foster scanner. The Foster scanner, or Variable Path scanner, is a type of **radar** system that produces a narrow beam that rapidly scans an area in front of it. Foster scanners were widely used in post-World War II radar systems used for artillery and mortar spotting. Modern radars in this role normally use **electronic scanning** in place of a Foster scanner for this purpose.

There were a number of contenders for satisfaction of the MES, principally from the UK and USA but also from France and Sweden. After staffing around a list of potential equipment it became obvious that for language, particularly with documentation and training plus the fact that they were used by armies close to our own the choice narrowed to two claimants the UK Green Archer and the USA AN/TPQ-36. This did not exclude all others but certainly gave me something to work on when building the defence budget bid spreadsheet.

Green Archer

A/N KPQ 36

The contenders were: The Green Archer, also called Radar, Field Artillery, No 8, was a widely used British mortar locating radar operating in the X band using a Foster scanner. Developed by EMI after an experimental model by the Royal Radar Establishment, it was in British service from 1962 until 1975 with the Royal Artillery, and

The AN/TPQ-36 Weapon Locating System a mobile radar system developed by Hughes Aircraft Company and manufactured by Northrop Grumman and Thales Raytheon Systems.

The A/N TPQ 36 is weapon-locating radar, designed to detect and track incoming artillery guns and howitzers, mortars and rocket fire to determine the point of origin for counterbattery fire

Despite the appeal of the Green Archer's ready availability and proven capacity and lower base price it was old technology and did not meet with the artillery requirement for guns, howitzers, mortars and rocket launchers location. This meant that we would build our bid around the A/N TPQ-36.

The MES budget as mentioned before was calculated on the Principal Item Price (PIP) and was the sum of PIP times the number of equipment ordered for operational, training, reserves and repair pool use, the base cost.

Further, a percentage of base cost was allocated for each support element, training, parts, documentation etc. The sum of all these was the budget and a line item on the five year rolling programme, where the plan is broadly fixed for the first 2 years and stays more indicative for later ones, providing an overview of the structure and coverage of the evaluation policy. The plan needs to be annually updated.

All budgets were calculated in $US and therefore dependent on the exchange rate, which in 1979 was in our favour with AS$1=US$1.13. By the time the MES had moved to becoming a budget item in 1983, the exchange rate had reversed (AS$1=US$0.88)thanks to Keating floating the dollar on Sunday 26 March 1983, one day after the election. This resulted in our order being reduced by a quarter and some fast staffing needed to keep the MES on track. Thank God, I had moved on by then.

Eventually, the A/N TPQ 36 was procured and was first deployed in the Australian army in 1987.

One of the principal differences between my Surveillance MES in 1978 and the Counter battery radars or Weapon Locating Systems MES in 1979 was the opportunity to accept hospitality from a wider range of gun-runners and one of the most regular were two men from Ericson electronics. One day they took me for lunch and then wanted to meet up after work for a drink. When I tried to beg off saying I had to pick up Carol it was arranged that she would have Keiran babysit and join us for the drink. The drink became dinner and champagne and cigars with dancing. A big night.

To repay them I picked them up after breakfast, and took them out to Royal Canberra Golf Club (RCGC). This was not to play golf but to satisfy their desire to have photos taken with some kangaroos. The easiest place to find kangaroos is on the RCGC early in the morning, and we found a large mob easy enough and they took many photos to take home with them.

******

The Canberra times one Saturday in mid-1979, had a small ad offering 'free to good home' a female Maltese about a year old and becoming available due to a marriage break up. Not a problem, I rang and the guy said there was a second dog also a male a little older and I agreed to drop around but told him we were only looking for a small female. Taking the kids with me, which was a tactical error, I drove over and there they were two little dogs who knew suckers when they saw them. To cut a long story, short despite my determination to have a female only, both Sasha and Tom found a good home. We found out to our dismay about a month or so later, that not only was Sasha not spayed but

185

Tom was a healthy little breeder when they mated and Sasha was pregnant.

On 27 August 1979 Lord Louis Mountbatten was murdered by the IRA and Sasha gave birth to three males and one female puppy. We named them: Lord, Louis, Mountbatten and Edwina in memory of a part of the Royal family that I **had always admired.**

The dogs were soon snapped up and sold but for reasons I cannot recall, but might have something to do with Keiran, we kept Edwina. So here we were with three Maltese terriers, two spayed females, once bitten etc. and Tom.

Time floated by and soon we were that is approaching the end of 1979 the end date for this posting.

Now I could have sat back, and waited for the system to make a decision and cop it sweet, but I decided to see what was available and make a bid. A mistake as it turned out

But it was to be compounded by another more significant one a month or so later when given a choice of posting I chose poorly. The SO1 Postings was a friend and he found three options for me, an SO2 Electronics job in Melbourne, basically the same job I had before going to UK, which was the one I probably would have got if I'd said nothing, the 2IC of 4 Base Workshop Battalion in Bandiana, Victoria or a command job as OC Puckapunyal Workshop Company at Puckapunyal (Pucka) in Victoria. Ego caused me to opt for the command job without taking into account who would be my immediate superior. So I made a rod for my own back and handed it to a two faced sod and gave him an excuse to beat me with it. Some seven years earlier he had used a national serviceman 2Lt who happened to have a PhD in maths to help him, i.e. do for him, a Dip Maths. The 2Lt nominally worked for me so I knew what was going on and stupidly made it known to others; how to make an enemy for life.

But that was in the near future first I had to finish 1979 and this include another 'bush week with the school. Here I made a discovery that, in the minds of the 20 something year old girl teachers, I was a source of some sought after fashion items 'jungle greens'. They did not care much about the body in them but liked the gear. Maybe Che Guevara was making a comeback but I handed out two sets of my early, 30 inch waist greens, and saw them being worn regularly whenever I picked up the kids from school.

The kids had enjoyed the school and we anticipated no dramas when the end of year results was announced. The boys had sailed through with the odd comment but nothing serious, but Keiran, who had a small speech impediment and stuttered a little at times, was determined by her teacher to be slow and would not go up with her class. We said words like 'dash it' or something bovine that rhymes with that and decided to ignore the asinine ACT School's recommendation and sort out the problem, if there was one, in Victoria.(See

Chapter Eightteen – Puckapunyal

We continued on, with our posting order now finalised, and a date set for me to take over as Officer Commanding, Puckapunyal Workshop Company of 3 December 1979. I visited for a week in November and spent the time arranging MQ and school for the kids. The boys were easy as they went straight into Puckapunyal State School. With regards to Keiran I sought out private schools in the area and there was a Catholic Girls school, Sacred Heart, in Seymour the nearest town. I had a chat with the headmistress and she was only too happy to accept Keiran, and recommended a local speech therapist that could possibly help.

Now as a property owner I had, for the first time, to arrange to rent our Rivett house and arrange occupancy post our removal to the MQ in Pucka. The removal was not a problem but we needed to figure out how we were all, the family and dogs, going to travel to Pucka. Our car was only a Ford Escort, so although the distance to drive was short it was lucky that the kids and dogs were small.

Finding a MQ in Pucka was easy as there was a three bedroom house in Milne Bay Close that was designated 'For OC PWC' so we liaised with the chap I was replacing, and his wife, to complete the hand-over/take-over of the Workshop and the MQ at the same time.

A month before we were to move I was offered a most important position, best man at Max Farrow's wedding to Jean Ridge to be held at the old Mint in William Street Melbourne, on 3 January 1980. It appeared as if all our ducks were lining up with us being posted about an hour or so north of the wedding site with plenty of time to settle in to the MQ beforehand. It was a lovely day and was finished off with a reception at a Lebanese restaurant afterwards.

Max and Jean moved into Max's house in Miller Street, Richmond, that he had bought after selling the flat at Hawthorn that the 'odd couple' had shared and Carol and I would often make the trip down from Pucka to spend a night or so with them. Jean could speak Greek due to a previous relationship and Max who has a remarkable ear for language was soon equally adept. It was a treat to go out with them in Richmond and sit with them as they chatted with the restaurant owners and were treated like family. Over the years that have followed they have expanded their language ability and can now chat with restaurateurs in Turkish, Vietnamese and Thai as well.

Puckapunyal army camp was, and is, a major military centre that incorporates a number of units and schools that have access to a large free fire range that can be used by all arms for live firing exercises. During national service (NS) it was the site of 2 Recruiting Training Battalion (2RTB) and trained about half of the national servicemen. It is located about 17.5 kms from the town of Seymour and 120 kms from Melbourne, just off the Hume Highway.

In 1980-81 it was the home of:

- Headquarters, Puckapunyal area with a full colonel, Commander, and the camp coord staff that ran the camp allocating MQs, cutting the lawns and maintaining the area in general. They were also responsible for the golf course, gym, camp theatre, and the area swimming pool; these latter facilities were left over from NS days.
- 1 Armoured Regiment which was the home of the Leopard tank fleet.
- The School of Armour that trained all ranks of the Royal Australian Armoured corps and was the home of the Armoured Corps Museum
- School of Army Transport that trained all ranks of the Royal Australian Corps of Transport.
- 3 Camp Hospital that was a 50 bed hospital complete with attached dental unit to look after the health of the soldiers.
- School of Army Catering that trained all ranks of the Army Catering Corps.
- Puckapunyal Workshop Company, the Raeme field repair facility supporting the local units.
- 21 Construction Squadron RAE a divisional unit based in Pucka Area.
- The range supervisor and his staff that coordinated the use of the range and maintained it.
- 3 Base Ordnance Depot (3BOD)had a reserve pool fleet of vehicles in Seymour and an Ammunition depot in nearby Monegeetta, and
- a RAAOC Supply company in Pucka.

Milne bay Close was the last street of the Officers' MQ area and we shared it with one Lt Col Colin Toll, CI of the School of Transport, the OC 21 Construction, the 2IC of

School of Armour, the US exchange officer, a major, at the School of Armour, the Brit exchange officer, a major, at the School of Transport, the OC of the Supply company, and last but by no means least a Lieutenant Bob Fawkner who was the QM at range control, and more importantly to all who knew him ,was 'BOF' the cartoonist for the Army newspaper whose take on military life kept all of us amused and had for years.

Those living on the street lived up to its title and were a close knit group with Keiran at 13yo a very popular babysitter in an area where there was a lot of socialising She was a regular for Colin Toll whose position at the School of Transport required them to be heavily involved in the mess life. This provides some amusement when a new family arrived in the street and one of the young Toll girls was talking to the wife who was the mother of a new born baby boy. "What is his name?" she enquired.

"Kieron" said the proud mother.

"That is a girl's name, our babysitter is named Keiran, and she is a girl." said Miss Toll.

Reiner Frisch, the OC 21 Construction, Mal Boyle the OC Pucka Supply Company, Bob Fawkner and I were members of the Pucka Area Officers' Mess and Reiner and I would regularly meet after work and have a drink before heading home. The others in the Close all had their own messes and it would only be on the occasion of a mess dinner or mess party that we were invited to, that we might visit these messes. There were no restrictions but the norm was not to, unless invited.

Pucka was often decried by soldiers as a terrible posting but for us as a family it was our best posting ever. The commander Col Stan Maizey was a 'doer' he would see a need and make sure it was addressed, no questions asked, just do it.

The camp as mentioned before had many facilities and as far as Stan was concerned they were there to be used, by the soldiers and by their families. He was determined to remove the impression of a posting to Pucka as a punishment and make it a posting to be sought by soldiers, particularly, soldiers with families. When we arrived before Christmas and just as school was finishing we lucked into the best of what Pucka could offer. Naturally there was our mess Christmas party but we were also invited to the other messes and got to meet a number of officers and wives from around the camp. Stan made sure the camp gym and pool was open and staffed, by one of the three PTIs that were on staff, every day; and told the R Aust Sigs Sgt who was the projectionist at the Camp theatre that he wanted movies every afternoon for the kids as well as the usual three nights a week movies that he had been putting on. I am not sure what had been the norm but my impression was that the projectionist job had been only part-time under previous Commanders. I know we did not pay for the kids' movies but think there was a nominal fee for those at night.

The pool was of course a great thing for the kids, but to go without an adult they had to demonstrate that they could swim across the width of the pool safely. Keiran and James had learnt to swim during our trips to Bilinga and had no trouble receiving their passes to swim on their own but David had not achieved that yet, so set himself to prove he was a big boy and it was not long before he could easily swim across the pool. Freedom, for both the kids and Carol.

There was a nine hole 18 tee golf course in the middle of the camp with the seventh green right outside the Area Officers' Mess side door and it was a test of character to continue on and finish out the game at times instead of ducking in through the side door.

The workshop was spread between two sites with the main workshop area in Pucka itself and a second workshop

192

building just outside Seymour adjacent to the 3 BOD reserve 'B vehicle' park. For convenience the B vehicles were repaired at Seymour by civilian contractors, Seymour Vehicle Services, employed by them but working under military supervision in our facilities using our repair parts. In and out inspection of the vehicles for repair, i.e. inspection before and after repair remained an army responsibility.

In the military vehicles are divided into basically three classes:

- A vehicles which include all armoured vehicles tracked or wheeled, tanks APCs etc.
- B vehicles which include all non-armoured wheeled vehicles, trucks, utilities, cars etc., and
- C vehicles which include wheeled or tracked item of earth moving equipment, either self-propelled or towed; all self-mobile, self-steering, purpose-made cranes, cable laying ploughs; all industrial and agricultural tractors and rough terrain fork lift tractors.

The A and C were repaired in the main Pucka Workshop which had a 40 ton capacity mobile overhead crane compare to the smaller capacity crane at Seymour.

Access to the workshop was via a tank road designed to bear the weight of a metal tracked Centurion and therefore more than adequate for the much lighter and rubber-padded Leopard. The tank road lead to, both the School of Armour and museum and 1Armoured Regt, and out to the range.

To conduct field repair on the TFCS, an electronic and infrared (EIR) workshop was being built when I marched in and my first task was to ensure it met all the requirements we had identified during the maintenance evaluation in 1976. Principal among these were the clean rooms necessary to open the sight and remove, repair and/or replace the sight, optics and laser range finder modules. Once repair was completed the sight and laser rangefinder had to be calibrated and spec tested. To accomplish this, two concrete columns were to be sited in the workshop area and to ensure accuracy they had to be absolutely stable and surveyed in. A target was to be located and fixed a set distance away for calibration purposes.

Given this had been my raison d'etre for so long only a few years earlier I naturally spent time each week checking on progress. The clean rooms went in and were pressure tested and passed and I was watching with interest the fitting of the test equipment and noted how they located and built the sight testing columns on the SE side of the building on surveyed points on a concrete slab. The walls and sight tubes were to be installed later because the contractor thought it easier and more accurate to site the columns then enclose them rather than build a room and site the columns inside. It was not my task to tell the contractor how to do his job that responsibility rested with a Royal Australian Engineer major who was the project officer.

The sight tubes referred to previously were round 1.5 metre diameter concrete pipes set into the outer wall and pointed at the target. They looked like horizontally mounted culverts sticking out through the wall. The reason for this was to provide a protective sleeve around the laser beam when it was being used to range on the target and reduce the chance of dangerous scatter. It was part of the significant OHS restrictions on using lasers in workshops, or where unprotected persons may suffer eye damage. For another example of the concern with OHS, all of us using or exposed to lasers had to have a preliminary eye test as a base reference, then after your time with the lasers ended, a second test to determine any damage.

Everything was going well until I returned from my monthly meeting in Melbourne and found that in my absence, only two days, the contractor had built the wall and inserted the sight tubes and was very pleased with his efforts.

On inspection I could tell the sight tubes were perfectly positioned to ensure the laser beam would pass through the centre of the tube and there was no risk of scatter. There was one small problem, however, he had built the tubes so they pierced the wall at right angles, which made for a very neat looking arrangement but the target, which was a large water tank on a hill, off at an angle, not at right angles to the wall, and if we had lased through the tubes as currently built we would be aiming directly at the MQ area and risked blinding half of the Pucka residents. When I rang the RAE project officer to pass on the good news it ruined his day completely and he was soon on site at the workshop with steam coming out both ears as he saw for himself the problem. It was a simple fix, rip it all down and start again this time using as the line of sight, not neat right angles to the wall and parallel to the ground, but angled about 40 degrees to the right and slightly elevated to frame the water tank target. "OH! If all life's problems were so easy to remedy," he muses.

Sport at Pucka was a big part of life as it is in most Army camps. The workshop fielded teams in cricket, Aust rules football and athletics in inter-unit completion with the larger units having the advantage naturally except in Australian rules where the workshop was very competitive and did very well. My role was spectator and supporter and managed to see most games.

Rugby in Victoria was a minor sport but the services army navy and air force all played under the common banner of Combines Services rugby Club. I attended the pre-season meeting representing Pucka Area and it was at that meeting that we prepared the breakup of teams for the year ahead. The teams were regionally based with First grade, Third grade and Colts from Melbourne city, Second grade from Puckapunyal, Third grade and Fourth grade from the Mornington peninsula, with teams coming from OCS and AAS, in Third grade and from OCS and School of Signals in Fourth grade.

Because the All Blacks were playing Victoria that year on 7 June 1980, as part of their Bledisloe cup tour, Combined Services were asked to field a composite side against a Victoria under 23 Side as a curtain raiser and I was asked to be the Manager. The selectors chose a side from mainly First grade in Melbourne and the Third grade sides at OCS and AAS. My job was to try and arrange transport for the players selected to train together, at AAS given it was most central to all and to ensure all players, equipment, medical staff etc. were available where and when needed . The game was played at Olympic Park in Melbourne and we lost to a much better side. The all Blacks thrashed Victoria 45-6, but lost the Bledisloe cup when beaten 2-1 in the tests. They also drew with NSW 13-13 but lost to Qld 9-3

Our Second grade games were played mainly in Melbourne so my first responsibility always, as manager, was the logistic problem of ensuring we got there and back. A very experienced RAAC warrant Officer was the coach and we, he and I, came to be a good management team. Unfortunately, for both of us the exigencies of service life denuded our rugby team at times and we could seldom play our best side.

In fact I used to take my boots along and played a few games when we were short I found, however, that I must have forgotten to pack my speed when I left the UK because I was unable to do the things I expected when a gap opened.

One game we were so short that we could only field ten men and pack a front row of three men, with me as hooker. (?)This was before the rule that required experienced props to pack down or else it became Golden Oldies rules where

"OUR BOSS"

you could not challenge and the team putting in had to win their own ball. Still we won as many as we lost and when everyone was available it was a team to be reckoned with. This was to be shown the following year when the team was more consistent with most of our better players available and we went through undefeated. Twice I had to fill in but there were still enough ability in the side to win despite me. Everyone who played received a winner's medal and the team made sure I got mine at the end of year function. The workshop digger's club decide to have a gentle shot so asked Bob Falkner, BOF, to do a special sketch to mark the occasion.

They presented it to me at the Diggers' club just after the grand final and I was quite taken aback.

The y had not finished though, because then I received a plaque with a Datsun dipstick made very short by taking a piece out of the middle.. My staff car which I drove exclusively was a Datsun and I had been lax when checking the oil and it was found to be very low when taken into the service station for periodic service. It was a serious error, and I probably would have been annoyed with one of the drivers if they had done it, so I was politely booted up the arse by my senior Artificer Sergeant Major when he presented me my 'award 'in front of the assembled workshop that night. The message was plain 'you obviously do not need the full dipstick as you do not check the depth often enough'. The mounted dipstick was inscribed 'Presented to the Dipstick of the Year'. I copped it sweet and it stayed on my office wall for all to see for the rest of my time in the army. It is still out in the shed. See BOF'S Cartoon above.

****

For twelve months of my two year in Pucka I was the PMC of the Area Officers' Mess which meant that Carol and I were regularly providing or enjoying hospitality. Saran Maizey as commander was on the same hospitality round about and all four of us became very good friends with Carol helping Jeanette with the Ladies Committee, and Stan and I finding we had similar interests with drinking and playing cards high up on the list.

As mentioned before the School of Catering was at Puckapunyal one of the least onerous and indeed most pleasant tasks was to attend lunch at the school when a course was marching out.

The courses included basic cooks and stewards, Corporal trade tests and supervisor caterers and each time there final trade test was to prepare a lunch or a cocktail party for about thirty or more guests. The Masons and the Maizeys were always four of them. It did not hurt that the OC of the School lived close to us and we often had the odd BBQ.

****

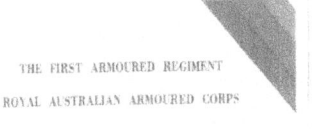

THE FIRST ARMOURED REGIMENT

ROYAL AUSTRALIAN ARMOURED CORPS

CEREMONIAL PARADE

ON THE OCCASION OF THE PRESENTATION
OF
THE STANDARD
BY
HIS ROYAL HIGHNESS THE PRINCE OF WALES

TUESDAY 21 APRIL 1981

His Royal Highness The Prince of Wales
Colonel-in-Chief,
Royal Australian Armoured Corps

Official program of Ceremonial parade for presentation
of Standard

On 21 April 1981, the Colonel-in-Chief of the RAAC,
HRH the Prince of Wales, presented 1 Armoured Regiment
with its first Regimental Standard. A parade was held on the
regiment's parade ground and there was a lunch afterwards
in the Officers' Mess. Carol and I were official guests so I
was in full ceremonial dress including Sword and Sam
Browne and sat with the other guests watching a fine display
of pomp and ceremony befitting the occasion.

STANDARDS AND GUIDONS

From official program of Ceremonial parade for
presentation of Standard

With the parade completed all official guests made
their way to the 1 Armd Regt Officers' Mess where we were
to join HRH for lunch. When we entered the mess I had to
take a toilet break and park my sword and Sam Browne after
accomplishing this I was to make my way to the dining
room. Now protocol states that all should be in the room
prior to the arrival of the official guest, HRH. I had been a bit
lax and found Himself closer to the door than I, so picked up
my pace with Carol close alongside. We still wonder what
would have occurred if he had not been stopped by someone
for undoubtedly we would have collided at the doorway with
me being the larger probably putting him on his royal back
side. As it was we crept in before him hoping to be unseen.
Prince Charles, we are on first name terms, is not a big man
but is most gracious, and charmed everyone as he strolled
around and shook everyone's hands.

Stan had seen my almost treasonous assault on the
future king and never let me forget it. Carol and I were to

meet Prince Charles and Diana in Hobart some years later but I did not raise the matter!

****

My career was about to come to a grinding halt, however, when the two-faced colonel who was my immediate superior gave me a 'not ready for promotion' assessment on my annual confidential report. I was in shock at receiving it as at no time had he called me in to give me a booting or made even the mildest indication of being unhappy with my performance. I should have redressed immediately but was so annoyed I said "stuff it" and did nothing. The problem was that I was in the frame for promotion and he knew it so he had taken his opportunity to thwart any chance I had of making Lt Col. What if I had redressed?

Chapter Nineteen - Return to AHQ

The rest of the year was a less than auspicious one and the final irony occurred when I was posted to a SO2 Major's position but being paid **Higher Duties Allowance** (HDA) for the vacant SO1 Lt Col position in Canberra, with a start date January 1982. The position was to be the Maintenance Engineering Agency's (MEA) eyes and ears in Canberra. There were two warrant officers and I, with another major's position also vacant, and our task was to ensure that the maintenance aspects of equipment procurement were not overlooked.

About mid-year I had to front the MS bosses to assess my readiness for promotion but given that last report it was not looking good. On reflection I did not do myself justice as I did not fight for my career but was so depressed by it all that I sat stunned and let it all happen, 'lay back and think of England'. The resulting 'passed over no promotion' outcome was not a surprise, but I was about to make a significant discovery that affected the progress of the procurement of the replacement B vehicle fleet, not that it would counteract the damage already done to my career. Raeme has the responsibility to recover damaged or broken down B Vehicles from the battlefield and/or in normal, peacetime, use. Previously, given the current fleet based on International 3 and 5 ton vehicles and the heavy haulage REOs, we had access to two primary recovery vehicles that were used by specialist Recovery Mechanics usually in a dedicated workshop environment and looked after by a team of RAEME trade qualified mechanics, serviced and cleaned more than most other vehicles:

A 5 ton twin boom crane, mounted on a 5 ton International chassis where each boom could independently lift 2.5 ton and 5 ton when used together as a single crane, It's off road capability was incredible and was fitted with the class leading Holmes Twin Boom recovery unit and was second to none in capability and flexibility of task applicability, and

The M543 wrecker. A rotating, telescoping, and elevating hydraulic boom by Gar Wood that could lift a maximum of 20,000lb (9,100kg).. Although the truck was not meant to carry a load, the boom could support 7,000lb (3,200kg) when towing. They had 20,000lb (9,100kg) front and 45,000lb (20,000kg) rear winches, outriggers, boom braces, chocks, block and tackle, oxygen-acetylene torches, and other automotive tools, the larger capacity heavy recovery crane based on a Reo chassis could lift 25 ton.

The replacement B Vehicle fleet was to comprise a fleet of 4tonne Mercedes vehicles called MOGs and heavy haulage Macks. Now the old recovery vehicles could handle the MOGs but only the aging M543s were capable of lifting and/or recovering the Macks.

The RACT project officer happened to have an office not far from me and as it was my habit to roam around and have a chat with the various project officers every now and again, so I visited him one day to listen in on progress and see if there were any concerns.

He had proudly displayed a collection of photos of all the new fleet showing the various chassis configurations including everything from cargo carriers to RA Sigs Shelters and Raeme Workshop vehicles. My attention was drawn however to what was not there, a heavy capacity wheeled recovery vehicle to replace the M543s. When I asked the question it was apparent that RACT were going to mould their budget to give more prime equipment and everything else was secondary. It had been assumed that if a MACK broke down they would use civilian tow truck operators.

Good thinking in 1982 prior to Desert Storm etc. and if your concept of vehicle use was bounded by Melbourne and Townsville. I wrote a paper and sent it to MEA HQ in Melbourne and made my point at the next meeting in Melbourne about a week later.

Needless to say the paper had caused some angst and there had been much wailing and gnashing of teeth but the reality was that the carefully balanced project had just been given a destabilizing nudge and provision had to be and was made for Raeme heavy recovery vehicles in the buy.

Midyear I was to get an opportunity to review a new piece of kit that the US services intended to bring on-line in the near future and it was with fascination that I flew in a helicopter as it navigated from RAAF Fairbairn to a beacon on top of Mt Ainslie using a little black box developed by Collins-Rockwell; it was my first view of a GPS and we could only sit and observe with awe as the operator counted down the feet to go before we were hovering directly over the beacon. This was all top secret and we were made very aware of that fact. The US did release the system for commercial and recreational use sometime later but at that time this bit of cutting edge navigation gear was too precious to allow out into the world. When they did finally release it they withheld access to some of the ephemeris data so it was less accurate than **mil-spec** but fine for general use.

****

By this time Col Dobie had left the system and had a small IT business in Kingston and had become involved in marathon running with a group from the Hash House Harriers, three SF officers and their WRAAC WO2 Clerk, Jan (I think!). They still did Hash runs and I enjoyed them so, although I had never been a distance runner, I took pleasure in their company and began to run part of their training distances as well, although at my own pace. The training regime was straightforward, every lunchtime it was a lap of the lake about 4.5 kms and I gradually found myself jogging the complete way even if gently. Then on the weekend they, the real runners, would run longer distances 10 then 15 then 20 kms out towards Tidbinbilla. My job was a one man support team and used to drive the route and position Bottles of water about every 5 km and drive along the route they were running making sure there was no problem. The goal of these keen runners was to compete in the 1982 Big M Marathon in Melbourne in September as an Army team. I became enthusiastic about this and offered my services as manager.

With milk drink Big M sponsoring it, Melbourne staged its first marathon in 1978 on a course that began in Frankston, headed up Nepean Highway and finished outside the Melbourne Town Hall. On 3 October 1982, the course started in Frankston, but used Beach Rd, finishing at the Arts Centre. Over 2000 runners took part in the inaugural event with most catching a special early morning train to Frankston. The race quickly went through a boom period with more than 4500 entrants in 1982.

Graeme Arnold, a RA Sigs captain was the best runner of the group followed by Col and the other two guys who were on a par, then Jan, with me the pedestrian. Graeme was the first runner I ever saw run with a radio. This was 1982 and it was a small transistor that he carried in his left hand. I saw this as the answer to my problem of allowing my brain to keep telling my body that is was hurting by distracting it with music and bought a Sony radio with earpieces some time later.

Given that 4 October was a public holiday in ACT, Labour Day, we decided to apply for leave from Friday 1 October and travel as an official team. This meant we received travel allowance and air tickets, covering the period 1- 4 October inclusive. As Manager I arranged our flights,

205

hire car and accommodation for three nights at a pub at Brighton on Beach road near the Brighton Beach railway station.

To allow for a cool down after the race I made contact with a swimming pool near the pub and arranged access, post-race, so we could have a swim and a stretch.

We left after lunch on Friday and settled into our accommodation in time to unpack and go for a run along the path heading towards Sandringham about a10 km round trip. To show willing I ran half way out and waited until they returned and came back with them.

Saturday was a rest day with a long walk followed by a train into the city to watch the early showing of Rocky 3. Thus it was with the strong beat of 'Eye of the Tiger' that we arose early and I drove the runners out to the start line at Frankston where we all dropped off bags that would be ferried in to the finish. Having seen them off I drove to the corner of Bay Road, Sandringham and Beach road, just over half way and cheered them through. Graeme was leading our group with the other guys some minutes behind and Jan about 20 minutes further back. Seeing her go past I drove in and parked the car at the pub and waited in my running gear for them all. When Jan came to me she was still running but feeling the strain so I dropped in alongside her and we jogged the remaining 12 km together. I think that was my first run over 10 km and it was indicative of the slower pace we were running that I was able to keep Jan going and not allow her to stop and walk. I think she finished in about 4 hours 30 minutes and was met with hugs and kisses by the rest of the team as she crossed the line. I did not run across the finish line but ducked around the side and joined in the celebrations. Graeme had broken the 3 hour barrier, which is the fun runner's goal time, 2 hour 48 minutes from memory, but the others were about 3hours 15 to 3 hours 30 minutes but finished feeling proud of their efforts. I was hooked and hoped that one day I would feel the same rush. We collected our bags and took the train back to the pub then we all went to the nearby, previously booked, pool and had a swim and consumed a few beers.

It was a very merry group that had dinner and drinks at the pub then in the morning drove to Tullamarine, returned the hire car, and flew back to Canberra..